INTRODUCTORY.

THIS little work has been prepared chiefly as a sequel to my "Elementary Geology," especially Section XI. But it may also be useful to all who desire in the easiest manner to get a general knowledge of the Geology of the Globe; for, though the text be brief, yet the Maps teach more, by a few moments' inspection, than many pages of letter-press. The long delay in the publication of this work, which was promised in the eighth edition of my Elementary Geology, in 1847, has been unavoidable.

E. H.

AMHERST COLLEGE, JANUARY 1, 1853.

OUTLINE

OF THE

GEOLOGY OF THE GLOBE,

AND OF THE

UNITED STATES IN PARTICULAR:

WITH

TWO GEOLOGICAL MAPS,

AND

SKETCHES OF CHARACTERISTIC AMERICAN FOSSILS.

BY

EDWARD HITCHCOCK, D.D., LL.D.,

PRESIDENT OF AMHERST COLLEGE, AND PROFESSOR OF NATURAL THEOLOGY AND GEOLOGY.

BOSTON:
PHILLIPS, SAMPSON & COMPANY.
1853.

Stereotyped by
HOBART & ROBBINS,
NEW ENGLAND TYPE AND STEREOTYPE FOUNDERY.
BOSTON.

THE GEOLOGY OF THE GLOBE.

A DESCRIPTION of the geology of different countries, in a connected manner, has sometimes been called *Geological Geography*, or *Geographical Geology*.

To attempt to give the geology of the whole globe may seem little better than conjecture, especially when wide districts exist whose geography, even, is not known. The attempt, however, has been made, especially by M. Boué, a distinguished French geologist, under the auspices of the Geological Society of France; and this little treatise will attempt to reproduce the results, in a form better adapted to popular use. The labors of American geologists will enable me to make some corrections of the geology of North America; and the geological explorations constantly going on in all parts of the world will furnish other facts, to correct the imperfections of theory. The outlines of this map I have copied from Johnston's Physical Atlas, Edinburgh edition of 1848, introducing those alterations which new researches enable me to do.

It can be only the larger groups of rocks which can be represented on a geological map of the globe. They amount only to six.

1. *Hypozoic and Metamorphic Strata, with Granite, Syenite and some Porphyries.*

I distinguish between hypozoic and metamorphic strata; for, though I admit the existence of the latter over wide districts, yet they are not, of course, beneath all the fossiliferous rocks, but may be merely their prolongation; and I believe, with Sir Henry de la Beche, that "beneath all the fossiliferous rocks there are mica and chlorite slates, quartz rocks, crystalline limestones, gneiss, hornblende and other rocks, of earlier production. These may indeed be merely altered or metamorphosed detrital and chemical deposits of earlier times, and possibly organic remains may be discovered in them; but, until this shall happen, it seems desirable to keep them asunder, for the convenience of showing previous accumulations." — (*The Geological Observer, Am. Ed.*, 1851, p. 32.)

2. *Primary Fossiliferous Strata to the top of the Carboniferous.*

3. *Secondary Strata.*

4. *Tertiary Strata.*

5. *Alluvial Deposits.*

6. *Volcanoes and Igneous Rocks of the Alluvial and Tertiary Periods.*

At the bottom of the map the principal mountains of the globe are represented, with their longitude corresponding to that on the map, their heights according to a scale, and stated also in figures, and their geology by colors, as shown by the tablets.

It will be seen that, with unimportant exceptions, all the surface of the map is colored; nor can I doubt that it is done approximately correct. Probably, however, only the predominant rock is given in many instances; and there may be small patches of other rocks in the same region not represented; and the exact limits of the several rocks are not exhibited in many cases. That the reader may judge of the real value of the maps appended, I shall present a brief outline of the means by which they have been colored.

The first and most important means is ACTUAL OBSERVATION. Almost all the islands and peninsulas of the globe, with very much of the sides of all the continents, have been explored by men acquainted enough with geology to give at least the leading features of the regions they have traversed. Nearly the whole of Europe has been carefully examined by able geologists, and the several formations traced out in much greater detail than is given on these maps. The same is true of the Atlantic portion of the United States of North America, and of parts of South America; and hence I have given a second map of the United States, on which much smaller and more numerous groups of rocks are represented than on the general map, and of which a fuller account will be given in the sequel. The first grand object, then, has been to use all known facts respecting the geology of different countries in the construction of these maps.

The next resort is to THEORY. It would be strange if the vast amount of facts made known by geology should not enable its cultivators to deduce, to some extent, the unknown from the known; in other words, to ascertain certain laws which have regulated the distribution of the different rocks. I shall now proceed briefly to enumerate the most important of these laws, as they have been presented by M. Boué, referring for their more full elucidation to his original memoir.*

1. *A knowledge of the Geography of a country is a most important means of deciding upon its Geology.*

Example. — Knowing the general geological constitution of the mountains of Peru and Mexico, and the physical structure of the Rocky Mountains, we might very safely infer in the latter a central band of crystalline or hypozoic rocks.

2. The prolongation of mountain chains, and even of continents, will generally be found of the same geological character as the main part of the chains or continents.

Examples. — The formations of Turkey in Europe

---

* *Memoire a l'appui d'un essai de Carte geologique du globe terrestre; presente le* 22 *Septembre*, 1843, *a la reünion des naturalists a Allemagne a Gratz.* — See *Bulletin de la Societie Geologique de France.*

are continued into Asia Minor ; those of the Crimea into the Caucasus. The principal volcanic chain of Central Africa passes over from N. W. to S. E., into the southern part of Arabia. The island of Cuba is prolonged into Jamaica. The formations of Norway pass into Spitzbergen, and those of the Ural Mountains into Nova Zembla.

This principle sheds light upon the geological character of the sea-bed, and, indeed, we already possess imperfect charts of the bottoms of the European seas.

3. The identity of direction in certain mountain chains, as well as their parallelism, throw some light on their geology.

Two chains on different parallels of latitude may run in the same direction as measured by the compass, and yet, if prolonged, they would intersect, and hence they would not be parallel.

Examples of chains identical in direction and constitution, but not parallel. — A good one occurs in the east and west chains of Europe and Asia : *e. g.*, the principal Alps, the chains in the south of Transylvania, the Balkan, Taurus, Indo-Cush, Kuanlun, Himalaya, and the Mountains of Heaven. Across middle Africa is probably a schisto-granitoid range, flanked by secondary and tertiary deposits. N. E. and S. W. chains occur in Ireland, Scotland, Scandinavia, Finland, Norway, Bohemia, south of Spain, Ala Dagh, Ak Dagh, Altai ; in southern China and western Hindostan ; between Herat and Candahar; the principal chain of the Alleghanies and Ozark Mountains; in Colombia, Brazil, New Holland and New Zealand. These show the same ancient schists, the same primary, fossiliferous, secondary and tertiary deposits, and the same porphyries, diorites and metallic formations.

Examples of Parallel Chains. — Compare the Hebrides with Scotland, Sweden with Finland, the Italian and Greek peninsulas with Morocco and the south of Spain, the two borders of the Red Sea with those of the Persian Gulf, and west Greenland with the neighboring American continent.

The older the formations, the more likely is this principle to hold true.

4. A knowledge of the Hydrography of a country aids as much in determining its geology as does its Orography, — that is, a description of its mountains.

The *Potamography*, or a description of the rivers in different formations, is strikingly dissimilar.

a In the oldest slate-rocks, such as talcose and mica slates and gneiss, the streams are characterized by their quantity of water, their bifurcations, and their winding course, as in the central Alps, in Brazil, in China, Tonquin and Cochin China.

b In calcareous countries the streams are less winding, the valleys are often dry, and the rivers are enclosed between steep escarpments, as in western Arabia.

c Knowing the character of the deltas of some rivers, we may infer that of others.

d Whenever we find large plains along rivers, higher than their deltas, we infer the presence of tertiary formations, because they exist in all such cases that have been examined. Hence we feel sure of their existence along the Senegal, Gambia, Orange, Euphrates, Indus, Ganges, Burrampooter, and the Blue and Yellow rivers in China, and the great rivers of South America.

e Sudden changes in the courses of rivers are a strong indication of changes in the character of the formations, and of the movements they have undergone.

Examples. — The Rhine at Basle and Bingen; the Rhone at Lyons; the Elbe in Bohemia and near Dresden; the Danube at Ratisbon; also the Euphrates, the Nile, the Oronoco and the St. Francis, of Brazil. This principle applies also to the Niger in Africa.

The bend in the Rio del Norte, in New Mexico, says M. Boué, indicates that the crystalline or hypozoic schists are flanked by older fossiliferous rocks; for all the bends of the great rivers going east from the Rocky Mountains occur at the junction of different formations; *e.g.*, the Missouri and the Mackenzie; thence he infers that there are metamorphic and hypozoic rocks on the east side of the Rocky Mountains, since all the great rivers there make bends as they come out from the mountains. American explorers there, however, have found the newer fossiliferous rocks more abundant than the older.

On the western slope of the Rocky Mountains Boué infers the existence of similar palæozoic rocks, from the bends in the Columbia and the Rio Colorado; and these, he supposes, connected with those of Russian America. In drawing these inferences, he regards the Rocky Mountains as the Urals of North America. This is certainly a sagacious suggestion, though it would more forcibly apply to the Sierra Nevada range, and its continuation northward, the Wind River Mountains. These run nearly parallel to the Rocky Mountains, and correspond almost exactly to the Urals in the character of their rocks, which are highly metamorphic and auriferous. Very probably they are of silurian age; but travellers have thus far rarely found in them silurian fossils. The strata correspond rather to the older crystalline rocks, which underlie the country from the east side of the Rocky Mountains nearly to the Pacific. Frequently, indeed, and over wide districts, newer rocks, such as the tertiary, the cretaceous, and the carboniferous, occur. But, in the present state of our knowledge, I have thought it safest to represent the whole of this country as older crystalline strata, except a strip along the Pacific, which I have colored as tertiary, especially from the testimony of Philip Tyson, Esq. (Report to Col. Abert, Ex. Doc. No. 37, of Thirty-first Congress, first session.) I have also marked a few other places, between Missouri and the Pacific, as underlaid by newer rocks ; but of their extent I know almost nothing. (See Prof. James Hall's report on specimens collected by Capt. Stansbury in his "Exploration of the Valley of Great Salt Lake in Utah," p. 401.) On both the accompanying maps I have omitted the broad tract of older palæozoic rocks, given on Boué's map, and have increased the amount of more recent volcanic rocks. But, after all, the geological character of the western part of this continent is so imperfectly known, that probably the attempt to exhibit it at all on a map will be thought unwise.

In South America, all the great rivers going eastward from the Andes make a curve as they enter the tertiary formations.

The existence of lakes in a country, their form and outlets, afford another means of determining its geology. If they are small, and lie upon the tops of mountains, they merely indicate that the waters part from thence. If in a plain, or in valleys, which have no outlet nor marked border, they may have resulted from a shrinkage in a secondary or tertiary arenaceo-gypseous formation. But if surrounded by heights, they may be genuine craters, as the lakes Ooromiah and Van in Asia, and Nicaragua in America. This sort of excavation, however, may have resulted from an eruption of granite, porphyry or serpentine, as well as of trachyte. Of the former an example occurs at Cairngorm in Scotland, and at Tatna in Hungary.

2

The width of lakes in the crystalline schists is usually much less than their length, and they have strait bays, with islands. They are masses of water, filling longitudinal or transversal clefts in the older stratified rocks. Examples in Scotland, Scandinavia and Finland.

The lakes of the older palæozoic formations have an oblong form, a waving outline, gently-sloping shores, islands, &c. They abound in the northern hemisphere, in America, Europe and Asia. Boué evidently reckons the great lakes of this country among the number. But their form does not well correspond to his description; and the largest of them, Lake Superior, is chiefly bordered by the oldest crystalline rocks.

Such lakes are usually produced by original movements of the rocks, or by erosion. Such rocks, particularly the calcareous, have been more favorable than any others to the production of cavities, by aqueous action, the explosion of gases, or other agencies.

The often deep lakes of the oolitic and cretaceous formations sometimes occupy spots broken in, resembling craters, as at Joux and Prespa, in Turkey in Europe. Others are merely cavities, resulting from the swallowing up of rivers, whose channels cannot carry away all the waters. Examples at Topolias in Greece, and those of Carniola, Dalmatia, &c.

Other lakes are but the filling up of deep clefts in calcareous rocks, as in the Alps and China.

5. Deserts furnish some, though not much, indication of the geology of a country.

Deserts are regions almost destitute of water, without trees, and with only a stinted vegetation. They occupy spots, for the most part, which at no very remote period have been gulfs, straits and mediterraneans, though some deserts, as in Arabia, are rocky, and have long been above the waters. Neither must deserts be confounded with steppes, which are usually in part tertiary. Some of those in Russia are mainly composed of a peculiar formation, deposited in brackish water during the tertiary period, when the Caspian Sea had an enormous extension beyond its present limits.

If there be lakes in a desert country, without outlets, the probability is that the soil is calcareous.

6. A general description of the aspects of mountains and their escarpments is sometimes useful to indicate their geology.

Notable differences in the height of different chains would lead to the suspicion that they belong to different ages and different formations. If a high ridge is bordered by lower ones and by plains, we may suspect a central nucleus of ancient formations, flanked by newer ones. *Examples.* — The Alps, the Himmalayas, the Mountains of Thibet, of India and of Tartary. Comparing the Uralian Mountains with those of the two Americas, we should suspect deposits of secondary and tertiary age, at least here and there, on the oriental slope of the Rocky Mountains.

Conical hills, with a crater, point out volcanoes. A series of peaks, like a saw, indicate dolomite; ridges broken down into holes, here and there, are apt to be calcareous; triangular points show quartziferous schists; needles are usually formed of the crystalline schists; unusual foldings indicate serpentine, or trachyte; pyramidal forms are frequently phonolites. Narrow and black walls show basalt, trachyte, or trap.

7. In a few cases, facts derived from botanical and zoological geography aid in understanding geology. *Examples.* — Some plants and insects are peculiar to a salt soil, some plants are peculiar to a calcareous soil, others to a sandy soil; some to a soil derived from trap, others to soil from granite, serpentine, &c. But these nice adaptations have not yet been enough studied to be of much use in geology.

8. Better indications are afforded by simple minerals; for many of these are limited to a single formation, or have a narrow range vertically. A catalogue of minerals and metals from China, for instance, afforded much light concerning its geology.

9. Organic remains furnish most important means of ascertaining the geology of a country.

Agassiz settled the geological position of a formation in the Alps, that had resisted every other mode of determining its age, by the character of its fishes.

Even the microscopic polythalamia have sometimes decided the character of rocks over wide regions ; an example of which is given in my Elementary Geology, p. 32, eighth edition. As a general fact, each formation has its peculiar and characteristic fossils ; and if we see only the fossils, therefore, we can tell to what rocks they belong. And these can sometimes be obtained from regions which no geologist has penetrated.

10. Erratic blocks, or boulders, furnish another means of determining the geology of unknown regions.

Examples.— If a geologist were to cross the northern part of the United States, and carefully notice the foreign boulders, he would be able to give a general account of the geology of the Canadas, even though for a great part of the distance wide lakes intervene. So he might pass around Scandinavia, commencing south of the Baltic, and going easterly into Russia, and from the boulders form a tolerable idea of the rocks in the first-named region.

11. The knowledge acquired by examining the structures and mode of filling up of some geological basins would afford important aid in deciding upon the geology of others similarly situated ; as, for instance, those of the Blue and Yellow rivers, in China, and of rivers in the north of Asia.

12. Man and his works, his distribution and degree of civilization, throw light upon the geology of the countries which he inhabits.

Examples. — If we inquire concerning such a country as Asia Minor, of low hills or plains, surrounded by chains of mountains of crystalline slates, we might be in doubt whether these hills were secondary or tertiary. The latter we might infer, if many roads are made over these hills; and on this principle Colonel Hauslab anticipated, in 1832, the recent discovery of these formations there. Traces of the ancient canal across the Isthmus of Suez are another example where the works of man enable us to conjecture something of the geology. The existence of a canal between Nankin and Pekin is another example.

The number of cities and villages increases almost in a geometrical ratio, from the oldest to the newest formations. Almost all the great cities are upon tertiary or alluvial strata.

Mines often determine the population and civilization. *Examples.* — Cornwall and the Hartz.

There is a connection between the civilization of nations and the configuration of continents. For instance, chains of mountains running east and west establish a greater difference between nations than those running north and south; also between the fauna and flora. The reason of this principle lies chiefly in the greater difference of climate between different parallels of latitude than between different parallels of longitude.

Hence it is said that almost all mixtures of natious and languages have occurred along north and south lines.

This same principle has greatly modified wars and conquests, since it is always easier to carry on war in an east and west direction than in one north and south.

The geology of the slopes of north and south chains is said to be usually similar ; but of east and west chains, dissimilar. North and South America afford examples of the first case, and the Alps of the second.

The formations on different continents are far more alike on the same than on different zones. *Examples.* — The north of America and of Europe and Asia ; also, the pointed extremities of America and Africa.

When ignorant of the geography of a country, we can resort only to analogies of form and the character of their borders. *Examples.* — Southern Africa resembles in form English Hindostan, each surrounded by chains of mountains. They have a similar potamography, a like elevation, similar plateaux and terraces, and the chains that mark their outlines are geologically similar. Now, as in Hindostan the crystalline slates and igneous rocks predominate, so they probably do in Africa. We know something from observation of the extremities of the triangle, and of six or seven points in their side ; and have some knowledge of certain metals from their interior. In like manner, New Holland may be compared with South America, and from the geological structure of the latter that of the former may be conjectured.

Such are the main principles by which, when observation failed him, M. Boué has been guided in the construction of the accompanying map of the world. That they have a foundation in truth, no geologist will doubt. The main difficulty probably lies in skilfully applying them. No man, perhaps, is better qualified to apply them than this distinguished savan. Nor is this the first of such efforts; for he was quite successful, some years since, in preparing a similar map of Europe; that is, subsequent examination shows that he was rarely mistaken. We cannot doubt but that a good deal of reliance can be placed upon this his more extended map. If his ambition has led him to strain some of his principles too far, in order to embrace all the globe, we can well pardon him. Enough reliance can undoubtedly be placed in this map to make it very valuable as a popular view of the geology of the globe,— an object of great interest, especially for teachers of the young.

Turning our attention now to the map thus elaborated, we can (in connection with observed facts) deduce some principles of a more general character respecting the geology of the globe, which will be of interest.

The seven parts of the world, recognized by geographers (North and South America, Europe and

Asia united, Africa, Australasia, Antarctic Continent and the ocean), are modified by the geologist as follows:

1. North America, embracing Mexico and Guatemala.

2. South America.

3. Africa, excepting its northern part, along the Mediterranean, or that part north of the Desert of Sahara.

4. Asia, with its great peninsulas, of which Europe is one, and the Barbary States in Africa another, and these are the principal.

5. Australasia and its isles.

6. Antarctic Continent.

7. The Ocean.

We have here six independent continental masses, with their appendices; each one presenting enormous masses of hypozoic or crystalline rocks, forming the skeleton; namely, two in North America, one in South America, two in Europe and Asia, one in Africa, one in New Holland, and one in the Antarctic continent, probably.

The six geological continents are now, or recently were, surrounded by seas, save the Isthmus of Panama. The Desert of Sahara was once (not long since, geologically speaking) occupied by an extension of the Mediterranean, the Barbary States forming an island.

The form of North and South America appears to be the type of all the other continents, if we unite

Africa to Europe, and New Holland to Asia. This renders it probable that their form is not accidental, but resulted from the mode in which internal forces acted.

The Arctic and Antarctic regions seem characterized by the rarity of active volcanoes, and the predominance of crystalline schists and granitoid rocks; towards the North Pole, however, embracing the coal formation. The patches of tertiary there are very recent.

The tertiary and secondary, associated with intermediate formations, more or less abound between the polar circle and the tropic of Cancer. The proper coal measures seem mainly concentrated within the same limits.

Volcanoes, as well as igneous rocks generally, though found on all parts of the globe, abound between the tropics.

Ancient igneous rocks, as well as volcanoes, are arranged somewhat circularly, like the mountains of of the moon. Examples occur in Bohemia and Ceylon.

The trachytes are chiefly abundant between the equator and the polar circle; basalt, between the latter and the tropic of Cancer, particularly in islands.

The relief of the surface has sometimes been much changed by modern volcanic action. For example, the Oocus, or Gihon, formerly ran into the Caspian Sea, but now empties into the Aral, in consequence of porphyritic or trachytic eruptions in the Bay of Kula, which shut up its ancient *debouchure.*

The fossils in the recent tertiary strata between the tropics, do not correspond to those now living there. Hence, since that period there has been a total change of climate all over the globe.

On going from the pole towards the equator, the resemblance between the genera and species of fossils and those now living increases. The exact identity, however, even in species, is rare, though some infusoria of the chalk appear to be analogues of living species.

The Geography of the World at different Geological Periods.

Geology is able to show with considerable accuracy what portions of our present continents were above, and what portions below, the waters of the ocean, at the different periods of the earth's pre-Adamic history, and thus to give an approximate view of its geography. Any continents or islands, however, which might then have been above the waters, and which subsequently sunk beneath them, cannot be described.

During the tertiary period, the Mediterranean covered the desert of Sahara, lower Egypt, and a part of Arabia. By looking at the accompanying map, and noticing those parts of Europe and Asia colored as tertiary, it will be seen that a large part of those regions was at that time under water. Recent researches have shown that an immense region around the Black, Caspian and Aral Seas, and extending on the east side of the Uralian Mountains, through Siberia, to the North Sea, was, during the latter part of the tertiary period, covered by brackish water, such as now fills the Caspian. In other words, that sea, during a long period, had that immense extent, and was not directly connected with any ocean.

It is probable that during the tertiary period the water of the ocean flowed through the valley of the St. Lawrence, so as to separate the Alleghanies from the hypozoic region around Hudson's Bay, as well as from the Rocky Mountains, forming, in fact, three large islands of North America.

Still more recently, however, even since the drift period, careful observations and measurements, of late, have convinced me that the waters of the ocean have stood from two thousand to three thousand feet above their present level in New England, and have either gradually been drained off, or the continent has been raised. Probably the waters have stood at a still higher level recently; but I speak only of an elevation where I have actually found sea-beaches and terraces, which the ocean only could have produced.

The state of the globe during the secondary period may be learnt approximately by supposing all those portions of the map colored as alluvial, tertiary and secondary, to be covered by the ocean. The tertiary seas at that time must have been in free communication with one another, and have been separated during the elevations of the secondary period.

During the period in which the palæozoic rocks were in a course of deposition, America was divided into six large islands, — three in North and three in South America. The Ozark Mountains also formed a small island or bank, in advance of the long island of the Rocky Mountains. In Africa, a large part of the continent formed a single island. In Europe, Scandinavia constituted the largest island, while the Urals formed another, and some six or eight smaller ones existed. In Asia, there were from three to five, though one of them was vastly the largest.

These statements probably present us with approximations only to the truth. It is not certain, for instance, that a region was above the waters during a particular period, because it is not now covered with the rocks peculiar to that period; for circumstances may have prevented depositions there, or they may have subsequently been denuded; and when we find the older crystalline rocks at a lower level now than some which still show marks of a mechanical origin, we must suppose the former beneath the waters while the latter were in a course of deposition, or else we must resort to the supposition of a subsequent depression of such spots. It has seemed to me that those who present us with maps of the world at the different geological periods have not sufficiently appreciated these difficulties.

We are now prepared to advance to a more detailed description of the earth's geology, though brevity will still be consulted. North America will be reserved to the last, because on that portion of the globe we wish to enter into more particulars. In order to give the geology in a connected manner, we must divide the continents geologically, rather than geographically; and in doing this we shall follow Boué's arrangement, for the most part, as given in his *Guide du Geologique-Voyageur.*

NORTHERN REGIONS GENERALLY.

These, as already intimated, are characterized by crystalline hypozoic slates and granites, particularly syenites; also by large deposits of other igneous rocks, not far from the shores of the sea. Some active volcanoes exist upon islands, such as Iceland and Jean Mayen. The tertiary formations almost entirely fail in these regions; yet some of the ancient tufas and trap rocks sometimes enclose masses of lignite or bituminous wood; but, as they are not accompanied by the leaves of trees, it is probable that this vegetable matter resulted from floating masses of wood. During the period of the coal formation, these regions were covered by plants of a tropical character, although now perpetual ice occupies the surface; and the change of climate appears to have taken place at the time when the subterranean lavas first found numerous orifices to the surface.

EUROPE.

In the central parts of Europe we find a complete series of secondary and tertiary rocks, with certain others of igneous origin. In southern Europe, along the Mediterranean, comprising the Pyrenees, the Alps and the Carpathians, the predominant deposits are a vast sub-Apennine or newer tertiary formation, a cretaceous system, a Carpathian sandstone (*gres des Carpathas*, a younger secondary rock), a peculiar oolitic or Jurassic system, and a peculiar trias, the coal formation being absent. Eruptions of diorite and serpentine have been frequent, and volcanoes, both extinct and active, occur on the sea-shore, or in the islands.

The north and north-east parts of Europe present us with the enormous development of older crystalline rocks, especially in Scandinavia and Russia. The ancient porphyry is abundant, especially in Sweden, where are wrought the finest rocks of this kind in Europe. In Sweden and Norway the rocks are highly metalliferous. The gold and silver mine at Kongsberg has been formerly regarded the richest in Europe, one mass of native silver having been found there weighing six hundred pounds. The iron mines of Sweden are in gneiss, and in 1839 yielded one hundred and twenty thousand tons. The copper mines of Fahlun alone yielded over five hundred tons, and cobalt is wrought in several places.

Silurian rocks (limestone and sandstone) occur in Sweden, under such circumstances as to indicate an ancient connection with those of Great Britain. Probably beneath the sea these formations extend from Scandinavia to Great Britain; and doubtless they are strewed over, as are those countries, with drift.

RUSSIA IN EUROPE.

The geology of the north of Europe, and especially of Russia, has been recently developed with great ability by Mr. Murchison, aided by Verneuil and Keyserling. — (*See the Geology of Russia in Europe and the Ural Mountains; two vols. quarto, London,* 1845.) In some portions of that vast empire these geologists find nearly all the series of known rocks. The tertiary strata are mostly confined to southern Russia, and belong, with one exception, to the older tertiaries. That exception is the Steppe, or Aralo-Caspian limestone, already alluded to, on another page, as having been deposited in a former brackish sea, which was an expansion of the present Aral and Caspian, and which must have been at least as large as the present Mediterranean. This rock contains a few peculiar shells, and some other remains, — such, for the most part, as are now found in the Caspian. It is divided by Murchison into a younger and an older Caspian formation, both of which, however, seem to be of the age of the recent tertiary deposits. He has marked the younger Caspian as extending along the east side of the Uralian Mountains to the mouth of the Obi river, on the Arctic Ocean.

The Jurassic or Oolitic System of Russia is found

in detached patches from the North Sea to the Caucasus; the cretaceous system is confined to the southern provinces. The trias, or upper new red sandstone, is wanting; but the Permian system is enormously developed, occupying a region more than twice as large as France, or more than four hundred thousand square miles. Indeed, we have here the type of this system, and the main evidence (not yet quite satisfactory to all geologists) that it belongs to the older palæozoic rather than the middle palæozoic rocks. There is also an immense development of the carboniferous system, one deposit of which, in the western part of the empire, is nine hundred miles long, and another, along the west base of the Uralian Mountains, twelve hundred miles long. The Devonian System, in the west and south-west part of the empire, is one thousand miles long. The silurian rocks, also, are well developed along the Baltic and in the Ural Mountains.

It is an interesting fact, that most of the deposits above described have scarcely been disturbed by volcanic action. The strata lie horizontal nearly, and to a great extent are scarcely consolidated. But when we go to the Ural Mountains the most decided marks of igneous action meet us; the whole chain has been upheaved and metamorphosed at a comparatively recent date, certainly not earlier than the carboniferous limestone; and the talcose, chloritic and quartzose slates there are, in fact, only metamorphic silurian, Devonian, and carboniferous shales and sandstones; while the granites, syenites, greenstones and serpentines, may not have been erupted till after the deposition of the tertiary strata, when it is supposed that the Urals were mainly elevated. Before that time Siberia formed a low continent, while Russia in Europe was beneath the waters.

The Uralian Mountains and Siberia have long been celebrated for the abundance and richness of the simple minerals and metals in them. Most of the gems, from the diamond to quartz, are found there, and the most valuable metals are gold, silver, copper, platinum and iron. The gold and platinum occur, almost without an exception, on the east side of the mountain, on a north and south zone, five or six degrees in length. Similar rocks also occur much further east, on some meridional spurs from the great Altai range; and it is said that in that part of Siberia is a region larger than France, every part of which seems to be more or less auriferous, — that is, it is covered by alluvia, from which gold may be washed,— and not only so, but all the subjacent solid rocks, if pounded up, yield gold. In fact, in the year 1843, the gold obtained in the eastern Siberian tracts amounted to more than eleven millions of dollars, raising the whole produce of the Russian empire to fifteen million dollars. More recently it has risen to about twenty million dollars. There is every reason also to believe that similar auriferous rocks extend southerly through Chinese Tartary, and into China, spreading, indeed, over the whole of northern Asia. Platinum, being in far less demand than gold, is not now as much explored in the Urals as formerly.

Mr. Murchison seems to have proved clearly that the rocks containing gold were impregnated by it at a quite recent period; namely, during some of the last eruptions of granite, syenite, greenstone and serpentine, which seem not to have taken place till after the deposition even of the tertiary strata, since these contain no gold. He states, also, as a general fact, that veins containing gold, that rise from the depths of the earth, are rich in that metal only towards their upper limit. Hence the greater part of this metal has been worn out of the rocks and is found in alluvium, and only one mine of gold in all Russia is wrought in the rocks. These facts are important, because the history of gold in Siberia is probably essentially its history in other parts of the globe.

Siberia has long been known as the most remarkable locality on the globe for containing the bones of the mammoth, and some other extinct quadrupeds. They occur in the same alluvia that contain the gold; and the banks of many of the rivers that flow into the Arctic Ocean, as well as some of the islands in that ocean, are full of them. The case of one carcass, found in a high bank of mud by Mr. Adams, not yet decayed, is well known. More recently others have been found in a similar condition, as well as one rhinoceros. The conclusion is inevitable, that while this auriferous alluvium was accumulating, probably in fresh-water lakes, these animals lived in great numbers on their shores; and that the last vertical movement of the Urals, by producing deluges and changing the climate, destroyed the race. Murchison, along with Humboldt and Lyell, thinks that no other cause of that change need be sought than the elevation of those northern regions a few hundred feet; and they quote the opinion of Professor Owen, founded on the structure of the mammoth's teeth, that this animal might have browsed, like the reindeer and the moose, in high northern latitudes. But we must bear in mind that the Siberian mammoths, although a striking case, were only one example of animals analogous to those now living in tropical regions, having formerly occupied the colder parts of Europe and America. If it be possible that one of these animals might have been acclimated in those northern regions, yet is it probable that they all were? Is the rule by which the comparative anatomist judges of the temperature of particular geological periods to be so easily set aside?

The phenomena of drift in Scandinavia and Russia are as interesting and instructive as in any part of the world, not excepting North America. It has long been known that the crystalline rocks of Scandinavia have been scattered southerly and southeasterly over the plains of Germany and Russia. But Mr. Murchison has marked out their limits upon his map, and these have been transferred to the map accompanying these remarks. Furthermore, it seems now to be ascertained, with a good degree of probability, from the researches of Böhtlingh, Siljeström and Nordenskiold, that Scandinavia was a centre of dispersion, from which boulders were thrown off eccentrically, towards the north in Lapland, towards the east and south-east in Russia, and towards the south in Germany, and perhaps towards the south-west into Great Britain, as is indicated by arrows upon the map. The greatest distance they have travelled is certainly not less than from seven hundred to eight hundred miles. In scarcely no case have they reached the Uralian Mountains, and Siberia is entirely free from them.

Admitting these facts, and every logical mind is forced to the conclusion that the origin of the drift of northern Europe must have had some connection with the elevation of Scandinavia from the ocean. Mr. Murchison supposes that the highest portions of that country were above the waters at the commencement of the drift period, and all the lower parts of Sweden, Russia and Germany, beneath them. Glaciers, he supposes, then existed on the Scandinavian mountains, from which icebergs were detached, loaded with the detritus. As that region was raised still higher, — at intervals, perhaps, — enormous waves of translation were rolled off *quaquaversally*, carrying with them masses of detritus, which grated over the sea-bottom, and produced the erosions and striæ, now so common. This was done while yet

Russia, and other regions where the phenomena of drift occur in northern Europe, were under water. Afterwards they were lifted up *en masse*, and in the same manner he supposes Scandinavia to have been elevated. The far-travelled boulders he supposes were transported by icebergs.

That water and ice were the agents which produced drift in all parts of the world, would now be generally admitted by geologists. But as to the particular mode in which they have acted, and the part each has played, there will be a diversity of opinion. Many, for instance, would doubt whether the most powerful wave of translation, impinging against a thick mass of detritus, would move it forward like a glacier, so that the lower portions would form parallel striæ upon the rocks. Would it not rather tear up the top first, and hurl forward successive portions in the most tumultuous manner? But this is not the place to raise objections to this theory, so ably defended by Mr. Murchison. Had he given ice more prominence in the work, it would have come nearer our own views; and yet, while we are fond of seeing others bring forward their ingenious views as to the *modus operandi* of ice and water, we would not be hasty to commit ourselves to any hypothesis on a subject concerning which a multitude of fascinating theories by the most gifted minds have proved as evanescent as the summer cloud with its rainbow hues.

In the southern part of Russia, and covering a

tract as large as an empire, there occurs an interesting deposit of black earth, called by the Russians *Tchornozem*, containing from six to seven per cent. of organic matter, and admirable for cultivation. It commences where the boulders and coarse gravel of the northern drift terminate, and in the opinion of Mr. Murchison was deposited in the ocean subsequent to the drift period, when the land was gradually rising from the water, and was mainly abraded from the black Jurassic limestone. This would, of course, bring the formation within the alluvial or recent period of geology. In this connection we are reminded of the *Regur*, or Indian cotton-soil, of the East Indies; of the *Pampean formation* of South America, and the prairie soil of North America, all of which had probably an analogous origin.

DENMARK.

Denmark is a flat country, underlaid by chalk and weald clay, over which is boulder clay, with huge Scandinavian blocks. The deposit appears to have been mostly made in more quiet waters than were generally concerned in the production of drift.

GREAT BRITAIN.

A large part of the north-west portion of Europe appears at no very remote period to have sunk beneath the waters, leaving Great Britain and a few smaller islands above them. A rise of six hundred feet would connect Great Britain with Scandinavia by wide plains.

If we consider Great Britain, the north of France and of Belgium, and a part of Westphalia, as constituting a geological province, it will be a region eminently distinguished for the development of the carboniferous system; also the oolitic and cretaceous systems, and the more ancient tertiaries. Trias is also common without *muschelkalk*, which is a peculiar German limestone, connected with the trias in that country.

Although this region contains ancient plutonic rocks, no igneous eruptions seem to have taken place there since the middle oolite period.

. Near Elgin, in Forfarshire, Scotland, the tracks of a chelonian, or tortoise, have been quite recently discovered on the old red sandstone. There also the skeleton of a Batrachian reptile has been disinterred, and described by Dr. Mantell under the name of *Telerpeton Elginense*. As these are the earliest examples of a vertebral air-breathing animal yet found on the globe, they possess great interest, and the figures A and B, Plate 1, represent both the tracks and the skeleton, the latter of the natural size.

No country on the globe has been so thoroughly explored, geologically, as Great Britain. Indeed, its formations have hitherto been regarded in a great measure as a type of the formations in other countries. And, indeed, they do not lead into much error, when so considered. The older crystalline rocks are most developed in Scotland, the north part of Ireland, and the west part of England. The secondary and tertiary rocks, so prolific of astonishing organic remains, occupy more especially the central and eastern parts of the latter country. Scotland is most remarkable for the drift phenomena; though the east part of England is strewed with detritus derived from Scandinavia. The mountains of Wales are mostly destitute of drift, and seem to have been the seat of former glaciers, which have left the marks of their descent through the valleys. Marine drift, however, occurs there to the height of twenty-three hundred feet.

No country has developed its mineral resources more fully than Great Britain. These are principally coal, iron, lead and salt. In England are five great coal deposits, and more than fifteen million tons are annually dug and consumed in England and Ireland, in which latter country are several deposits, as well as in Scotland. Iron also usually accompanies the coal, and annually one million three hundred and fifty thousand tons are manufactured. The whole amount of salt manufactured in England is about fifteen million bushels, two of the beds at Northwich being about sixty feet thick. Of lead, the amount smelted from the ore in 1828, of which twenty-nine thirtieths was obtained in England, amounted to ninety-eight million seven hundred thousand pounds; and this is said to be three times as much as is obtained in all the rest of Europe. The

4

amount of copper obtained from the mines of Great Britain annually, is twelve thousand tons; the amount of tin, three thousand six hundred tons; of zinc, two thousand tons; and of silver, six thousand pounds.

HOLLAND AND BELGIUM.

The greater part of Holland and Belgium is low and flat, — some part even below the sea, and defended by dykes. Yet some districts are mountainous, and in Belgium the strata are as remarkably inverted and folded as in the Appalachian chain of the United States; and their history has been ably developed by Professor Dumont.

Both anthracite and bituminous coal occur in Belgium in great abundance. Of this, in 1837, her two hundred and fifty coal mines produced one million six hundred thousand tons, and the iron mines produced annually one hundred and sixty-two thousand tons.

SWITZERLAND.

The Alps, the highest mountains of Europe, occupying nearly all Switzerland and a part of taly, and stretching away, in the Julian Alps, towards Turkey, form, as it were, a sort of geological skeleton for central and southern Europe. The central and highest parts of the chain consist of the older crystalline slates, with the older unstratified rocks; while on its flanks we find distinct ridges of newer rocks, such as new red sandstones, oolite, lias, greensand, chalk and tertiary strata, some of the latter lying from two thousand to four thousand feet above the ocean, and showing a comparatively recent elevation of some parts of these mountains. Over all these formations we find vast accumulations of drift, the phenomena of Scandinavia being here reproduced. It appears, too, that in general the blocks have been thrown outward from the axis of the principal chains, presenting us with another centre of dispersion. If we were to confine our attention to the drift of Switzerland and the adjoining regions, we might reasonably adopt the unmodified glacier theory as to its origin. Indeed, one meets there with the most decided evidence of the former wide extension of the glaciers.

In Switzerland the mining operations are hardly worth naming, showing us that the useful metals are not commonly most abundant in regions where the rocks are developed on the most gigantic scale. The iron mines of that country (including Piedmont, Sa voy, &c.) produce twenty-two thousand nine hundred and fifty tons ; the lead mines, three hundred and fifty tons ; and the silver mines, one thousand two hundred and fifty pounds. The salt mines of Bex, however, now produce from two millions two hundred and forty thousand, to two millions eight hundred thousand pounds.—(*Hand-Book of Switzerland*, p. 162.)

The phenomena of folded and inverted strata are well exhibited in the Alps on a stupendous scale, as in Belgium and the United States.

GERMANY.

Almost every known variety of rock occurs in Germany. Over the extensive plains of that country, including Prussia and Poland, are extensive tertiary deposits. Further south the country becomes mountainous, and all the older rocks are well developed. Its geology has been very thoroughly studied and described by the many able geologists who have lived and still live there.

The coal mines of Germany produce annually about one million tons; her salt mines, one hundred and fifty-seven thousand five hundred tons; her iron mines (Austria one hundred thousand, Prussia one hundred thousand, the Hartz Mountains seventy thousand), three hundred thousand tons; her lead mines, eighteen thousand five hundred tons; her silver mines, fifty-seven thousand pounds; her copper mines, six thousand tons; her mines of mercury, nine hundred tons; and her gold mines, three thousand three hundred pounds, mostly found in Austria.

FRANCE.

The type of German geology will answer essentially for France. Her tertiary rocks are abundant in not less than six basins; her secondary deposits, especially of chalk, extensive; her carboniferous rocks produce, from two hundred mines, not less than one million one hundred and fifty thousand tons of coal annually; some of her mountains contain the older crystalline slates, and some of them are extinct volcanoes, especially in Auvergne. In the tertiary and newer deposits occur numerous species of extinct quadrupeds, which have been described. mainly by Cuvier, in his great work, the *Ossemens fossiles*.

A splendid geological map of France has recently been published by MM. Brochant de Villiers, Elie de Beaumont and Dufrenoy, as the result of a geological survey of the country, by order of the government.

In 1839 the iron mines of France produced six hundred thousand tons, which is more than their average yield. About four hundred tons of lead and three thousand three hundred pounds of silver are annually smelted in that country.

SPAIN AND PORTUGAL.

It will be seen, by the map, that the Pyrenees, between France and Spain, eleven thousand feet high, are mostly of secondary palæozoic rocks. The older crystalline rocks occur most abundantly in other parts of the country. Near Olot, in Catalonia, is a region of extinct volcanoes, occupying twenty square leagues. In that country are the most productive quicksilver mines in the world, yielding annually one thousand seven hundred and fifty tons; twenty thousand five hundred tons of lead are also obtained, and fifteen thousand tons of iron. Gold was formerly washed from the sands. At Cardona, in the Pyrenees, is one of the most remarkable deposits of rock-salt on the globe, forming a mountain five hundred feet high.

The geological character of Portugal is so similar to that of Spain, that no additional remarks are necessary to explain it.

ITALY, ETC.

The Apennine region of Europe comprehends Italy, Dalmatia, Albania, Greece, Sicily, Sardinia, the south part of Spain, and parts of Africa and Syria. It is characterized by vast fossiliferous deposits, and by the sub-Apennine or post-pliocene formations.

TURKEY IN EUROPE.

In Turkey a portion is colored on the map as hypozoic schists, with granite. But, for the most part, these rocks are merely metamorphosed Silurian, Devonian or carboniferous rocks, and of course younger than the azoic schists of Sweden and Norway.

HUNGARY AND TRANSYLVANIA.

The Carpathian Mountains of Transylvania and Hungary are represented as mainly composed of secondary rocks. Those regions also abound in trachytes, with trachytic porphyries and tufas of more recent date than the tertiary period, and one solfatara at least exists in Transylvania.

Hungary is said to produce more gold and silver than any other country of Europe, except Russia.

The amount of gold has been stated at one thousand and fifty pounds annually, and forty-one thousand pounds of silver; but this is more than is now obtained. About ten thousand tons of iron, nineteen thousand tons of copper, and one thousand two hundred and twenty-five tons of lead, are also obtained. Large quantities of coal and salt are also dug from the earth.

SUMMARY OF THE METALS.

The value of the more useful and usual metals dug from the mines in the different countries of Europe, namely, iron, gold, silver, copper, lead, mercury, tin and zinc, is thus given by C. D'Orbigny and A. Gente, in their *Geologie Applique aux Arts*, etc., 1850.

Great Britain,	$88,000,000
Russia and Poland,	27,000,000
France,	26,000,000
Austria (including Hungary),	13,000,000
German Confederation,	12,000,000
Spain,	10,000,000
Sweden and Norway,	10,000,000
Prussia,	10,000,000
Belgium,	8,000,000
Tuscany,	3,000,000
Piedmont and Savoy,	2,000,000
Denmark,	2,000,000
	$211,000,000

Professor Ansted gives the following table of the amount of the above-named metals annually obtained in the different countries of Europe :

	Tin.	Cop'r.	Merc'y.	Zinc.	Lead.	Silver.	Gold.	
	Tons	Tons	Tons.	Tons.	Tons.	Lbs	Lbs.	Tons.
British Isles, ...	3600	12000		2000	46,000	6000		1,350,000
Russia and Poland, . .		3000		4000	600	38,500	12,000	300,000
France, ...		80			400	3,314		430,000
Austria, ...	20	3500	250	75	4500	42,500	3,250	70,500
German Confederacy, .	300	2000	650		8000	5,350	60	70,000
Spain, ...		25	1750	85	20,500			15,000
Sweden and Norway, .	65	1240		300	40	10,350	4	85,000
Prussia, ...		540		500	6000	10,000		70,000
Belgium, ...				1750	325	350		162,000
Tuscany, Elbe, &c., . .								24,000
Piedmont, Savoy, Switzerland, &c., . .					350	1250	13	22,750
Denmark, ...		700						12,000

ASIA.

In Asia we find at least two vast ranges of mountains corresponding to the Alps of Europe, and whose areas are composed of the oldest crystalline rocks; namely, the Altai and Himalaya chains. The first, however, may prove to be of the same character as the Uralian Mountains; that is, the crystalline schists may be metamorphic, and the unstratified rocks of comparatively recent date. But the Himalaya range appears to be of the earliest date, and to constitute what Boué calls *l'epine dorsale — the back-bone* — of Asia.

SIBERIA AND KAMTSCHATKA.

The vast region of Siberia is traversed by some mountains separated by wide plains. These mountains seem to be offsets from the Altai range, which forms the southern boundary, and runs nearly east and west.

On the accompanying map may be seen the geological structure of this vast region. The amount of alluvium along the shores of the Arctic Ocean, as well as on its eastern coasts, is greater than in any other part of the globe. The crystalline schists occupy the next largest space ; the older fossiliferous rocks, the next ; the tertiary, the next ; and the newer secondary the least space. Boué characterizes Siberia by "its large deposits of crystalline schists, its great amount of minerals and gems, its peculiar trias, the absence of the Jurassic and cretaceous systems, its tertiary strata in the vast steppes of Tartary, and abundance of salt, &c." "Eastern Siberia differs from western Siberia by its volcanic rocks, of various ages, its trachytes, its basalts, and even its volcanoes in Kamtschatka and the Aleutian Isles."

I have already given some account of the remarkable gold and platinum deposits of Siberia ; from which we can hardly avoid the conclusion that this cold and desolate region is likely to become an El Dorado, probably of much wider extent than that of California. In the Altai Mountains large amounts of silver are obtained, as well as gold. The entire amount of the former metal, in 1828, in all Siberia, was not less than one hundred and eighty-two tons. The gems found in that country are such as the diamond, topaz, emerald, beryl, onyx, lapis-lazuli, rubellite, avanturine, hyaline quartz, carnelian, chalcedony, agates, &c.

TARTARY AND THIBET.

Only the general geological character of that vast region of central Asia that goes by the name of Tartary is known. It will be seen by the map that the three great classes of rocks occupy vast areas there; namely, the older crystalline, the older fossiliferous and the tertiary. Volcanic mountains also occur there, such as Pechan, Houtcheou, Ouroumptsi, Kobok, and Aral-toube. They form the eastern prolongation of that wide belt of extinct and active volcanoes in Asia, which occupies not less than two thousand five hundred square leagues. Thibet is separated from Tartary by the Kuenlun Mountains, but not much can yet be said of its geology. Large quantities of borax are obtained from a lake in that country, which contains also common salt.

HINDOSTAN.

Hindostan is separated from Thibet by the Himalaya Mountains, the highest on the globe. The plains south of this range are underlaid by a sandstone of the mesozoic period, next succeeds clay slate, then mica slate, talcose slate, quartz rock, hornblende slate, and limestone. The highest part of the range is gneiss, traversed by granite veins. The mica slate is traversed by porphyry. Tertiary strata also occur at the foot of tha mountain, containing bones of the mastodon, elephant, hippopotamus, rhinoceros, elk, horse, deer, crocodile, gavial, sivatherium and monkey. The Jurassic and cretaceous systems are also widely developed in these mountains, as in the Alps of Europe, of which they, in fact, form the counterpart, or rather the prototype. Drift is said to exist at the base of the Himalayas, but it is chiefly the result of ancient glaciers.

In Peninsular India unstratified rocks are enormously developed, as granite, porphyry, and especially trap. In the porphyry and trap we have mines of tin and gold, with fine crystals of quartz, and the various zeolotic minerals, with chalcedony in its various forms. The annual value of carnelian exported from India, a few years since, was fifty thousand dollars. Pegu produces that most beautiful gem, the oriental ruby, or sapphire. The diamond is found in Hindostan, as well as corundum in all states. Topaz, zircon, tourmaline, garnets, jasper, agate, amethyst, chrysolite, &c., are common. In Burmah nearly one hundred thousand tons of petroleum are annually collected in wells. Coal and rock-salt also occur in India.

CEYLON.

Most of Ceylon is composed of the older crystalline rocks, gneiss being the predominant rock, with dolomite, quartz rock, granite, syenite and trap. Gems are numerous in the gneiss, such as amethyst, rose quartz, catseye, prase, topaz, garnet, pyrope, cinnamon-stone, zircon, spinel, sapphire, and corundum.

CHINA AND JAPAN.

China probably represents central Europe, with its rich coal-fields, its trias, its Jurassic system, its great tertiary basins, and its deposits of salt and petroleum.

In Japan, M. Siebold has described crystalline schists, granite, trachytes, or even lava and tertiary deposits. Volcanic rocks, indeed, are quite abundant.

EAST INDIAN ARCHIPELAGO.

In the centre of the semi-elliptic zone, that extends from the Philippine Isles to Barren Isle, and the Isle of Flora, we find crystalline schists with palæozoic rocks, as in Borneo and Celebes. In Banca occur gneiss and granite with tin. Sumatra, Java, Borneo, the Philippine and Molucca Isles, are volcanic. Four of the volcanoes in Sumatra are twelve thousand feet high. In the Isle of Sunda enormous masses of volcanic and trachytic rocks rest on those which are azoic, and are surrounded by tertiary and sub-Apennine beds.

The mineral treasures of this Archipelago are abundant. Gold occurs in almost all the islands, and the amount annually collected falls little short of three million dollars. In 1817 the island of Banca produced over two thousand tons of tin. Sulphur is found in purity in Java; and Borneo is celebrated for its diamonds, one of which, in the possession of a native prince, is valued at one million one hundred and ninety-four thousand and thirty dollars.

AUSTRALIA.

New Holland and Van Diemen's land present geological types analogous to those of the north of Europe, in the predominance of older crystalline rocks, in their encrinal limestones, coal formations, and the secondary sandstone and limestone of the north-west part of New Holland, and the tertiary rocks with lignites and the osseous breccias, and recent littoral deposits of its central parts. Numerous peculiar quadrupeds have been found fossil there, such as the kangaroo, dasyurus, hypsiprimnus, wombat, a rodent, a saurian and an elephant.

The recent discovery of gold in abundance in New Holland is a matter of deep interest. Rev. W. C. Clarke, ever since 1841, from personal examination, had been urging public attention in vain to the subject. While in England, Sir Roderick I. Murchison, from the resemblance of the Australian geology to that of the Ural Mountains, had done the same. At length men were convinced of the correctness of these conclusions, and for a year or two past they have rushed to this new locality, which threatens to rival even California. Mr. Clarke estimates the auriferous region in the Macquaire district alone to embrace seven hundred or eight hundred square miles. Already the shipments of gold to Europe are counted by tens of thousands of pounds, almost monthly. A single specimen was lately found weighing over forty-seven pounds. The amount produced by that country in 1852 is estimated at not less than five millions of dollars.

The rocks of the gold district of New Holland are the older azoic slates, traversed by veins of quartz. Mr. Clarke mentions the curious fact, that just ninety degrees west of the auriferous range in Australia we find the analogous range of the Urals, and just ninety degrees east the similar range of the Sierra Nevada; and all the ranges have a nearly north and south direction.

The remains of eleven species of extinct gigantic birds in the alluvium of New Zealand, found a few years ago, constitute the most remarkable zoological discovery of the present century. Professor Owen describes them under the name of Dionis (six species), Palapteryx (three species), and Aptornis (two species). They were wingless, mostly three-toed birds, varying in size from that of a turkey to that of a bird at least ten feet high. The natives call them Moas, and probably they existed within a few hundred years. Fragments of their eggs have been found, which indicate a size about equal to a man's head.

The discovery of these giants has taken away all improbability from the existence of congeric races as early as the new red sandstone period, as is indicated by their tracks in the Connecticut valley.

ARABIA, PALESTINE AND SYRIA.

In passing to the south-west part of Asia, we meet, along the shores of the Arabian and Red Seas, in Arabia, a vast deposit of crystalline azoic rocks, of which granite and syenite are frequent members. The huge pile of mountains called Mount Sinai are composed mainly of granite, or rather syenite, a variety of granite. Porphyry and greenstone exist in the same mountains, and the latter is found at Akaba, and ancient volcanic craters, according to Burckhardt. The valley of the Jordan, or rather the Wady el Arabah, extending from the Red Sea to Mount Lebanon, was probably a fissure, along which volcanic agency may have been more or less exerted, from time to time. It does not appear, however, that any lava of consequence occurs in this valley, till we reach the Sea of Gennesareth, in Palestine. Near Safet is an extinct volcano. Most of the rocks in Syria and Palestine, however, as well as those in Arabia, and extending into Egypt, are limestone, of the age of the chalk, and probably also of the oolite. Mount Lebanon, eleven thousand feet high, is mainly composed of the former rock, and in it abundant Polythalmia, Venus, Tellina, Terebra, Trochus, Dolium, Ostrea, Hippurite, Echinodermata, and fine fishes with homocercal tails. East of Palestine and Syria, wide tertiary deposits stretch away towards Persia.

ASIA MINOR.

The older crystalline rocks exist in Asia Minor; but more of the surface is occupied by secondary and tertiary rocks. This country has also been powerfully affected by volcanic action, as it lies directly in that volcanic zone extending from the west part of Europe through the greater part of Asia. The Katakekaumena, or Burnt District, near the ancient Philadelphia, is well known as abounding in lava and ancient craters. The vicinity of Smyrna exhibits also recent volcanic rocks. Further east, is the ancient Armenia, Georgia and Persia; and there we find marks of volcanic agency, particularly near Erzeroum, around the Lake Van, where occur vast trachytic deposits, at Elbrouz, Demavend, Orfa, Sindjar, and especially at Mount Ararat, which, although nearly eighteen thousand feet high, is composed of lava; and some contend that, within a few years, it has sent forth at least vapors, if not lava. There are also many thermal springs in Armenia and the north part of Persia, which indicates the proximity of volcanic action, though not necessarily very recent.

M. Dubois has recently made an examination of the Caucasian chain of mountains, extending from the Black to the Caspian Sea, and his results are very interesting. Although in lithological characters almost every variety of rock seems to exist in the central axis of that chain, yet in fact it is composed of Jurassic and liassic deposits, lifted up ten thousand feet, and highly metamorphosed, while tertiary formations stretch away from either flank over the wide plains. Dubois supposes that this chain has continued to be elevated even as late as historic times; and he suggests that one of these vertical movements, by draining large lakes, may have produced a deluge in the time of Noah, universal, at least so far as man was concerned. It is said, also, that another geologist finds a peculiar mud deposit in Armenia, which he suspects may have been left by the Noachian deluge. But, whether this may not be a continuation of the Tchornozem of Russia, already described, remains to be seen. These speculations of M. Dubois, however, are very interesting, and his views seem in part sustained by the fact that mud volcanoes still exist at both extremities of the Caucasus.

ANTARCTIC CONTINENT.

This inhospitable region, so recently discovered, and covered continually with snow and ice, presents no very attractive field of research for the geologist. But the French navigators obtained specimens of granite of different colors from the land, and the American exploring expedition found quartz, sandstone and conglomerate, on icebergs near the coast; and they describe certain mountains on the coast as having a black, volcanic aspect, from all which we may infer the presence of a variety of rocks and volcanoes, — the latter, indeed, were seen in action. Boué says that chains of crystalline schists occur there. The small circle upon the map will show what is known of the geography and geology of these regions.

POLYNESIA.

The vast group of islands known by this designation are wanting in the older rocks, and exhibit a foundation either of coral reefs or of lava. The former are low and level, the latter uneven and high. Mouna Roa, in the Sandwich Islands, is more than sixteen thousand feet high; and Mouna Kea, eighteen thousand four hundred feet. These islands, embracing more than four thousand square miles, are all volcanic, and Kilauea is the most remarkable active volcano on the globe, as a reference to my Elementary Geology, where it is described, will show. The Friendly Islands, the Gallipagos, the New Hebrides and Gambien Islands, are mostly volcanic; though such islands are usually fringed with coral reefs. Juan Fernandez is made up wholly of basaltic greenstone. The South Shetland and Orkney Islands contain primary rocks; but these belong rather to the Antarctic continent. The coral islands are such as Elizabeth Island, Whit-Sunday Island, Queen Charlotte's, Lagoon, Egremont, &c.

It was a former hypothesis that coral islands were formed by the polyparia building their reefs on the margin of volcanoes; but Mr. Darwin and others have shown the fallacy of this view; and that geologist suggests that many of those reefs are built upon islands that were gradually subsiding, so that the coral-builders were able to keep their structure just at the surface of the waters. Mr. James D. Dana has endeavored to show that a region of the Pacific Ocean, thirty degrees on each side of the equator, embracing not less than fifteen millions of square miles, has within a comparatively recent period undergone subsidence. If we draw a line nearly east-south-east from New Ireland, near New Guinea, just by Rotumah, the Society Islands, &c., all the islands north of it, with two or three exceptions, are purely coral islands, while those to the south of that line are generally basaltic. Hence Mr. Dana infers that the region to the north of said line has been undergoing a depression. — *Am. Jour. Science*, vol. XLV. p. 131.

AFRICA.

Of central Africa we know but little from actual observation. On the river Zaire, however, crystalline slates and granite predominate, as well as in the kingdom of Sackatoo and at the Cape of Good Hope. In the interior of this last colony there are secondary calcareous sandstones, with tertiary strata, yet the region is destitute of igneous rocks. On the Nile, however, and between that river and the Red Sea, extinct volcanoes are described. In Egypt and Ethiopia generally, however, we find crystalline schists with dioritic and syenitic eruptions, which are flanked by ancient secondary formations, and above these a cretaceous formation, with nummulites. Associated with a recent sandstone there exist, especially near Cairo, large quantities of silicified dicotyledonous trees. The Great Desert, Sahara, is represented on the map as tertiary. In Ethiopia is a lake sunk below the ocean, like the Dead Sea. Along the western coast gold is washed from the sands. A hundred years ago its annual amount was equal to two to three hundred thousand pounds sterling. The northern part of the continent, embracing Mount Atlas, is described as chiefly secondary and tertiary, with some deposits of ancient crystalline rocks.

A large part of the islands around Africa are volcanic, as the Azores, the Canaries, where Teneriffe rises to the height of twelve thousand feet, also the Cape de Verd Islands, St. Helena, and, in the Indian Ocean, Maurice and Bourbon. Madagascar is probably an ancient isle of hypozoic rocks, with recent deposits along its shores.

Very recently M. St. Hilaire has made known the existence of the eggs and bones of a bird in Madagascar, which must have lived not many centuries ago, and which was larger than the Dinoris giganteus of New Zealand. The egg was in diameter eight and a half by thirteen and a half inches, or about seventeen English pints in capacity, equal to six ostrich eggs, one hundred and forty-eight hen's eggs, or fifty thousand humming-bird's eggs. St. Hilaire gives this bird the name of *Æpyornis maxi-*

mus, and thinks it must have been at least twelve feet high.

AMERICA.

The western side of the continent is traversed by the longest, and, with one exception, the highest chain of mountains on the globe. In South America it is called the Andes, or Cordillera. On the Isthmus of Panama it sinks into a comparatively low ridge, which in Mexico and Guatemala rises into a high table land. In the Rocky Mountains beyond, we have an elevated prolongation of the same chains almost to the Arctic Ocean, and a parallel chain, the Sierra Nevada, quite near the coast. Parallel to the Atlantic we have the Appalachian or Alleghany Mountains, rarely more than six thousand feet high, and usually not more than from two to three thousand. Branches extend irregularly into Canada and around Hudson's Bay, and the mountains of the West Indies are perhaps their southern prolongation. South of the delta of the Orinoco we find numerous ridges in Guiana and Brazil.

On this continent we find also vast plains. That between the Appalachian and the Rocky Mountains is the largest in North America. On the Atlantic coast, south of New York, we have another. In South America we have the vast Pampas of Orinoco and La Plata.

SOUTH AMERICA.

South America may be described geologically

under five regions; namely, Brazil, the Guianas, Colombia, the Plateaux of the Andes, and southern Chili, with Patagonia.

Brazil seems to have formed a very ancient continent, composed almost entirely of crystalline schists, with gems and metals, such as iron, gold and platina. Quartzose rocks abound among them. The only eruptive masses are granite and diorite. This continent was formerly separated from the Andes on the west by wide straits, now occupied by tertiary formations. The secondary rocks of Brazil belong to the cretaceous and oolitic periods.

The Guianas correspond in their nature essentially with the ancient deposits of Brazil, which are separated from Brazil by the Amazon, and from Colombia by the Oronoco. In Colombia the crystalline schists are covered here and there by the coal formation, lying at great heights (sometimes eight thousand feet). We find there also the old red sandstone, a saliferous and gypsiferous trias, without muschelkalk, and deposits of the oolitic age and character; over which lie more recent deposits.

Peru is characterized by its immense plateaux of crystalline schists, flanked here and there by palæozoic and mesozoic arenaceous deposits. But its most striking character consists of immense domes of trachyte and ancient volcanoes. Large deposits of precious minerals, and vast quantities of salt and gypsum, also occur.

Chili is characterized by its palæozoic rocks, its

carboniferous strata, its enormous mountains of trachyte, its active volcanoes, and its recent littoral deposits. The phenomena of drift also occur there.

In the late work of Charles Darwin, entitled *Geological Observations on South America* (London, 1846), we have very interesting views of that country; and though perhaps not contrary to the views of Boué, above presented, yet they give us a more accurate idea of the geological structure of that great continent.

If I understand Mr. Darwin, his account of the distribution of gneiss, mica slate, talcose slate and clay slate, which he calls metamorphic schists, corresponds essentially with their representation upon the accompanying map. But, in respect to the newer formations, superimposed upon this ancient floor of the ocean, and the vertical movements which the continent has undergone, he gives us much new light. What he calls "the basal strata of the Cordillera" is a purplish or greenish porphyritic claystone conglomerate. The fragments, sometimes rounded and sometimes angular, and often six or eight inches in diameter, he supposes were ejected from volcanoes and then arranged by aqueous agency, so as now to exhibit a singular combination and intermingling of aqueous and igneous action. With the conglomerates alternate strata of true feldspathic porphyries, the whole having a thickness sometimes of six thousand to seven thousand feet; and its main mass lies along the central line of the Cordillera.

Another very interesting plutonic rock of great extent in the Cordillera is Andesite, consisting mainly of white albite, green hornblende, mica, chlorite and epidote, with grains of quartz. This occurs also along a large part of the axis of the Cordillera, and appears to have been erupted at a more recent date than the porphyries above named.

About the time when the claystone and greenstone porphyries had nearly ceased being erupted, a vast deposit began, which Mr. Darwin calls the gypseous formation, from the great amount of gypsum which it contains. The masses of that mineral found in the Cordillera exceed anything of a similar kind in any other known locality. Beds several hundreds of feet thick, much of it beautiful white gypsum, are interstratified with sandstones and shales to the thickness of seven thousand or eight thousand feet, and extending over vast districts. From the organic remains found in these rocks, they appear to belong in part to the cretaceous and in part to the oolitic formation. Hence Mr. Darwin denominates the whole the cretaceo-oolitic formation. It is found sometimes at the height of fourteen thousand or fifteen thousand feet above the ocean; higher, I believe, than these rocks occur anywhere else. During their deposition feldspathic lavas and other singular volcanic rocks were erupted, and it is supposed that the gypsum, or perhaps only its sulphuric acid, was thrown out from numerous submarine craters. While the deposition was going on, the continent, where it now appears, must have subsided several thousand feet.

The tertiary formations of South America of very early date, according to Mr. Darwin, are of enormous extent. Including the Pampean formation of tertiary age, according to M. d'Orbigny, and also the boulder formation, we have a continuous extent of surface of tertiary deposits, twenty-seven degrees of latitude, or nearly one thousand nine hundred miles, extending also across the continent. Their thickness is about eight hundred feet, and at the foot of the Cordillera they rise to the height of three thousand feet. Mr. Darwin has made it probable that they were deposited while the continent was slowly sinking, in the whole not less than seven hundred or eight hundred feet.

Mr. Darwin did not discover boulders connected with drift in South America till he had reached the fiftieth degree of south latitude. In ascending the Santa Cruz river, when one hundred geographical miles from the Atlantic and sixty-seven from the Cordillera, boulders began to show themselves, and were strewed over the surface in increasing numbers to the mountains, from whence they were doubtless derived. They were generally angular, and some of them as much as sixty feet in circumference. The plain on which they lay scattered was about one thousand four hundred feet above the ocean. In Terra del Fuego is another deposit of boulders, which were derived from ledges lying from sixty to one hundred and twenty miles to the south-west and west. On the western coast no boulders are seen till we reach the Island of Chiloe, lying between south latitude forty-one and forty-three degrees. There they are extraordinarily numerous, and were probably derived from the Cordillera on the continent, about forty miles distant. Mr. Darwin supposes, with good reason, that glaciers and icebergs must have been the agents employed in transporting all the erratic blocks of South America ; and since they seem to have been shed off east and west from the Cordilleras, the probability is that the vertical movements of those mountains must have been concerned in their dispersion.

Recently Sir Robert Schomburgh has described enormously large and far transported blocks of crystalline rocks as occurring in British Guiana, which lies nearly under the equator. This is an important fact, if there be no mistake about it, especially as geologists have been disposed (hastily, as I have always conceived) to confine the phenomena of drift to the colder regions of the globe. But it needs a man who is familiar with drift in such regions as Scandinavia, or New England, to be able to decide in all cases whether travelled blocks belong to the true boulder formation, or were the result of other agencies.

The post-pliocene or alluvial formations of South

America are very interesting. The quantity of sediment carried down her vast rivers has been immense. At the same time the continent has been gradually rising, so that we find vast plains, now hundreds and thousands of feet above the ocean, covered with deposits of gravel and mud which were formed in ancient estuaries and since the drift period. I reckon the Pampean formation in the post-pliocene, following the opinion of Mr. Darwin, although M. d'Orbigny refers it to the tertiary period. But the fact that it contains the bones of vast numbers of terrestrial quadrupeds, now extinct, mixed with species of estuary shells, still living along the coast, makes it probable that it was deposited since the drift period, especially as this seems to have been the case with the great North American extinct mammifers, as well as those of northern Asia.

The Pampean formation consists of a more or less dull-reddish, slightly indurated, argillaceous mud or earth, including sometimes concretions of marl, and passing into compact marly rock, the latter called by the inhabitants *Tosca rock.* This deposit is often as much as one hundred feet thick, and embraces numerous species of such shells as are now common on the coast, with the remains of estuary infusoria and a vast amount of extinct quadrupeds. It has been traced continuously nearly eight hundred miles in length, and from three hundred to four hundred in breadth. Hence it occupies an area at least as great, and probably twice or thrice as great, as that of

France, which is two hundred thousand square miles. The most probable theory of its origin is, that it was the former estuary of the La Plata, including some of the bottom of the adjoining sea. Into this estuary the bodies of the mammifers that lived upon the adjoining land were washed in such numbers that Mr. Darwin thinks a trench could nowhere be cut across the formation without intersecting a skeleton. These animals are very peculiar, yet formed on the same general type as the existing animals of South America. They were such as the mastodon, megatherium, megalonyx, scelidotherium, mylodon, holophactus, toxodon, macrauchenia, equus, ctenomys, hydrochænus, glyptodon, kerodon, canis and armadillo ; often several species .of each.

Mr. Darwin seems to have shown, very satisfactorily, that the South American continent has experienced powerful vertical movements in mass in comparatively recent times. During the tertiary period he thinks there was a subsidence of some seven hundred or eight hundred feet ; and since the drift period the elevatory movement has been going on in general so quietly as to be unobserved, but sometimes by starts. In southern Patagonia the coast has risen four hundred feet, and in La Plata one hundred feet, within the period of existing shells, though not of existing mammifers. At Santa Cruz (latitude fifty degrees south) the plains have been upraised one thousand four hundred feet since the boulders were spread over them. These elevatory movements have been interrupted several times by periods of repose, during which extensive denudations took place ; and the coast of Patagonia in one place to the height of nine hundred and fifty feet, and in another one thousand two hundred feet, is formed into eight great step-like gravel-capped plains, extending hundreds of miles on the same levels. These plains were formed by the denudations that took place on the coast by the ocean and rivers during these periods of rest. Along the Pacific Ocean these elevations have not been so equable. There the coast has risen, since Indian man inhabited it, at least eighty feet. So gentle have been these vertical movements that they have not produced the slightest disruption or disturbance in the tertiary or newer strata, throughout their vast extent.

If these vertical movements be admitted, we can hardly avoid looking to the volcanic agency that has so long been active all along the Cordillera as the cause. From the earliest times has matter been erupted along that extended line. First of all, the submarine lavas were poured forth, alternating with porphyritic conglomerates, by which the ancient ocean was filled up. During the cretaceo-oolitic period feldspathic streams and abundant mineral exhalations were erupted; then came basaltic lavas, while yet the sea washed the eastern foot of the Cordillera. Next we have the ancient tertiary eruption, and finally those successive outbursts that have left extinct craters, or trachytes, and other recent lavas, or still pour forth, from time to time, exhalations (solfataras), flames and lava. Mr. Darwin has marked not less than twenty-one craters along the Cordillera south of the twentieth degree of latitude, such as Tupungato, Maypu, Coquimbo, Guanahuaca, Villarica, St. Clemente and Aconcagua, the latter having attained the enormous height of twenty-three thousand feet above the sea. From the eruptions of these mountains being nearly simultaneous in many cases, we infer a connection at least of those several hundred miles apart, and probably even one or two thousand miles. Now, we have here a power sufficient, no doubt, to account for the vertical movements of the South American continent. But usually it has been supposed that volcanic eruptions, being paroxysmal, must produce sudden elevation or subsidence of the earth's crust. Mr. Darwin, however, thinks that this agency may operate slowly, and almost imperceptibly. He supposes a solid crust of only some twenty miles, and that some unknown cause produces an irregularity in the melted matter beneath, producing a sort of undulation of the solid crust, and an intrusion of lava into the fissures produced. When these fissures do not reach to the surface, earthquakes are the result, accompanied with vertical movements. But, if the matter enters a fissure already existing through the crust, a volcanic eruption will be the consequence. He thinks Beaumont's hypothesis of the shrinking of the internal nucleus, and the sinking down of the too large envelope, not to be a sufficient cause of the elevation of continents and mountain chains. Yet he does not seem to us to have shown its insufficiency.— (*See his Paper on the Connection of certain Volcanic Phenomena in South America.* — *Geol. Trans. Vol.* v. *second series*, p. 601.)

Prevost has recently endeavored to account for the inequalities of the earth's surface in general by the unequal contraction of its crust. This view has been drawn out more fully and illustrated by James D. Dana, Esq., very ingeniously, in the American Journal of Science, vols. II. and III., new series. He supposes that beneath our present oceans the crust contracted most, and that this caused it to sink down upon the fluid nucleus so as to form beds for the waters. This would in part drain the continents. But it would also by lateral pressure elevate them, and perhaps fold together the strata, as has been done the whole length of the United States. Indeed, he supposes that the Appalachian chain lies along the Atlantic coast, and the Rocky Mountains and Cordillera along the Pacific coast, because of the greater contraction of the crust of the globe beneath those oceans. These views seem to present us with a cause — whether adequate or not to explain the phenomena, let our readers judge — for the vertical movements of this continent, or rather for the disturbance of the melted nucleus of the globe.

South America has been celebrated, ever since its discovery, for the great amount of the precious metals found there. The more common metals, also, such as copper, abound. In Brazil, gold and its associated metals predominate, almost to the exclusion of all others. It is chiefly in the mountainous district of Minas, where metamorphic rocks of no very ancient date occur, that gold is found. Platinum is found with the gold ; and also iron and manganese, disseminated in stratified rocks. The whole series of rocks, gneiss, chlorite slate, and altered sandstones, have been penetrated throughout by metallic particles. The diamond also is found there, in a white quartzose matrix, probably the Itacolumite. When these diamond mines were first discovered they sent forth one thousand ounces annually, but at present only one hundred and eighty-three ounces.

Peru, including Bolivia, or Upper Peru, has been considered, until the recent discoveries in California, Australia and Russia, as the most remarkable region on the globe for the precious metals. Here occurs the celebrated silver mine, or mountain of Potosi, which is eighteen miles in circumference, and almost an entire mass of ore. In two hundred and twenty-five years it yielded no less than one billion six hundred and forty-seven millions nine hundred and one thousand and eighteen dollars. The veins occur in clay slate, regularly stratified ; the vein stone is quartz and carbonate of lime, and the principal ores, even in this mountain, are iron, galena and blende, the silver being very subordinate. One quicksilver mine in this country is seventy feet thick, and formerly yielded an immense amount. It is said that the silver veins present a fact not uncommon in argentiferous veins; namely, that, though very rich at the surface, the value diminishes rapidly downwards. The same is true of gold veins.

In other parts of the Cordillera valuable metallic deposits are common. They usually most abound along the lines of junction between the porphyries and the stratified rocks, especially limestone of the cretaceous age. The mines of Pasco alone yield one hundred and eighty thousand pounds of silver annually. Those of Chili (gold and silver) yield eight million five hundred thousand dollars annually, and one hundred copper mines exist. These metallic treasures are confined mostly to the chains of the Cordilleras, and extend through Colombo and Guatemala to Mexico. Indeed, in South America, as we have seen to be the case in other parts of the world, it is where plutonic agency has been most powerful in dislocating and metamorphosing the stratified deposits that metals and gems most abound.

WEST INDIES.

The islands of the West Indies appear to be the remains of a former continental surface, connecting North and South America; that is, from some cause the continent is depressed at this point, and it happens that a volcanic belt crosses these islands and the adjoining isthmus. There can hardly be a doubt that there is some connection between this volcanic belt and this depression. Indeed, it would be the effect of the greater contraction of the crust of the globe where volcanoes exist, and thus does Mr. J. D. Dana account for the facts in the case. — *Am. Jour. of Science, vol.* III., *new series, p.* 95.

In Cuba, Hayti and Jamaica, the mountains rise to the height of eight thousand to ten thousand feet. In Cuba the highest peak is mica slate, and through the secondary deposits of the lower regions project gneiss, granite and syenite. Gold, silver and copper, occur here ; and, in one instance, coal exists in a vein, which is almost an unique example, as it almost everywhere else occurs in seams or beds. In the south part of Hayti is a mountain of granite. Jamaica, in its most elevated part, is composed chiefly of palæozoic rocks and trap. Above these occur red sandstone, limestone marl and porphyry. In the eastern part of the Caribbean group, as Tobago, Barbadoes, Antigua, Bermuda, &c., tertiary limestones predominate. In Antigua is an interesting deposit of silicified dicotyledonous wood, of the tertiary period, specimens of which, when polished, form splendid agates.

In St. Croix the palæozoic rocks, with tertiary and recent limestones, predominate, but no igneous rocks are found there. In the recent marl, or limestone of Guadalupe, — it is an exceedingly hard rock, — two human skeletons have been found, one of which is now in the British Museum, and the other in the

Garden of Plants, at Paris. They were doubtless the remains of Caribs, the original inhabitants of this island ; and, as the rock is constantly forming on the coast, we need not suppose them of any great age.

The western parts of the Caribbean group are mostly volcanic, as Grenada, St. Vincent's, St. Lucia, Martinique, Dominica, Guadalupe, Montserrat, Nevis, St. Christopher's, and a part of Jamaica. Trachytic and basaltic rocks are common, even on those islands where there have been no eruptions within historic times. Trinidad lies near the South American continent, and is composed mainly of the older crystalline or azoic rocks. Here is the famous Pitch Lake, three miles in circumference. It is mainly covered by mineral pitch, or asphaltum, often hard enough for a man to walk over it. The thickness is unknown. As this substance is usually of vegetable origin, probably this immense mass was driven to the surface by volcanic agency from a coal deposit beneath.

NORTH AMERICA.

The numerous geological surveys that have been executed in North America within a few years past, by order of the different governments, enable me to give a more definite account of our geology than would have been possible a few years ago. For these surveys have furnished the materials for a good geological map of the United States and some of the British provinces ; and I have ventured to extend this map across the whole continent. Yet I wish it to be understood that west of Missouri and Iowa the coloring is intended to give only a general view of the geology, and probably in many parts it may be greatly erroneous. I have made several important modifications of Boué's delineations.

I shall now proceed briefly to describe the rocks of North America, beginning with those that are crystalline and usually considered as the oldest, but which the present state of geology shows us to have been produced at almost all periods of the earth's history.

HYPOZOIC AND METAMORPHIC STRATA, WITH GRANITE, SYENITE AND PORPHYRY.

The most highly crystalline of these rocks are usually associated in North America, and have an enormous development. The slates are mica slate, talcose slate, chlorite slate (a variety of talcose slate), quartz rock, hornblende slate, and gneiss. Through them all granite is protruded in the form of veins and irregular beds, or tubercular masses. This rock, however, usually occupies but a small extent at the surface, and frequently passes into syenite, or syenitic granite, by taking hornblende into its constitution, and losing most of its mica. The famous Quincy granite, in Massachusetts, is a good example of this rock.

This group of rocks is colored rose-red upon the map. It will be seen that they have an immense development in Canada, extending, in fact, to the Arctic Ocean. A narrow strip of more recent rocks in the valley of the St. Lawrence separates them from similar rocks in New England. From New England they stretch south-westerly, parallel to the coast, in a rather narrow belt, nearly to the Mississippi. On the west side of the continent they constitute the basis of the Rocky Mountains, and are represented by Boué as occurring around the Bay of California. I have reason to suspect, however, that when the Rocky Mountains and the western side of the continent have been more carefully examined, newer formations will be found superimposed upon these older slates, and their area be much circumscribed.

Associated with these rocks beds of saccharine limestone are not infrequent. Often metamorphic action has been so powerful upon this rock that its stratified arrangement is nearly or quite obliterated; as in the beds of it in the gneiss and syenite in the eastern part of Massachusetts; and in the northern part of New York, the limestone actually occurs in veins in the granite, according to Professor Emmons. This limestone is often charged with silica and magnesia; but frequently it forms beautiful white marble, and is burnt for quick-lime.

In the north-east part of New York, nearly the whole of the County of Essex, with its lofty mountains, is composed of hypersthene rock. This is composed mainly of Labrador feldspar, with an intermixture of that variety of pyroxene called hypersthene. It is, in fact, a variety of syenite, or syenitic granite. In that region it is associated with very large deposits of magnetic iron ore.

At the present day not a few geologists regard all the crystalline slates above described as metamorphic, and as having once been mechanical in their structure, and even abounding in organic remains. Internal heat, they suppose, has so nearly melted them down, that the particles could assume a crystalline arrangement, which has obliterated all traces of organic existence, although the planes of stratification and lamination generally remain. All geologists, indeed, agree that these rocks have been subject to powerful metamorphic action; yet some do not suppose that the oldest crystalline slates had an entirely mechanical origin, nor that they ever contained organic remains. But there are in our country extensive series of slates which all would denominate metamorphic. These usually lie between the most decided crystalline slates and the lowest fossiliferous rocks; for example, all along the west side of the Green and Hoosac Mountains of New England, and so through the Appalachian ranges to Alabama. Thinner deposits of similar character occur in many other places throughout the country. As to the principal deposit above named, there is a diversity of opinion concerning its age. Some, as the Professors Rogers, Hall, Adams, Logan and Sir Charles Lyell, regard the strata as the lower members of the silurian rocks metamorphosed; while Professor Emmons supposes them a distinct and lower formation, corresponding to the Cambrian rocks of Great Britain. He even points out peculiar organic remains from these rocks in New England and New York. This is not the place, however, to discuss this subject, though the first of these opinions seems to be generally received. It is sufficient here to state the fact that such a series of rocks, many thousand feet thick, and many hundred miles long, exists in our country. We cannot resist the conviction, however, that the evidence of their metamorphic character constantly increases. We are prepared, for instance, to point out a spot in Bernardston, Massachusetts, where the oldest non-fossiliferous clay slate of the Connecticut valley lies above encrinal limestones not older than the Onondago limestone of the New York survey. The rocks embraced in this metamorphic group consist of talcose, chloritic and argillaceous slates and quartz rock, with vast deposits of white, gray and black limestones, varying in texture from coarse granular to compact. We know of places, too, where mica slate and gneiss seem to be embraced in the series. The talcose slate is less highly crystalline than in the older series above described. Indeed, it passes insensibly into clay slate, and into hard, greenish compounds, that have been sometimes called serpentine. The porphyries of the United States are most usually associated with these metamorphic strata.

LOWER SILURIAN SYSTEM.

American geologists have described the rocks of this country idential with the upper and lower silurian of Europe, including two or three members of the Devonian group, under a variety of names. The New York geologists denominate the whole *the New York System,* which they subdivide into twenty-eight minor groups and four principal groups; while the Professors Rogers include the whole (with the Devonian and carboniferous systems), under the term *Appalachian Palæozoic Day,* subdivided into nine series, whose names have reference to the time of day, as the Matinal, Levant, Premedidial, &c., series. Without discussing the merits of any of these systems of classification, American or European, I shall describe them according to Professor James Hall, whom I consider better qualified to decide these questions than any man living, because he has had a better opportunity to study the fossils of these rocks than any other one. — *See his latest views, in Foster's and Whitney's Report.* Part II., p. 285.

The lower silurian system of Europe, according to Professor Hall, embraces the eight lower members of the Champlain division of the New York geologists, and the entire primal and matinal series of the Professors Rogers. The lowest formation is the Potsdam sandstone, three hundred feet thick; the next is the Calciferous sand-rock, three hundred feet; the next, the Chazy and Black River limestones, one hundred feet; the next, the Trenton limestone, four hundred feet; the next, the Utica slate, one hundred feet; and the highest, the Hudson River group, embracing the gray sandstone, seven hundred feet. Total thickness of the lower silurian rocks, one thousand seven hundred and twenty-five feet.

In lithological character the lowest part of this series is highly siliceous, much of the Potsdam sandstone bearing a close resemblance to quartz rock. In ascending, the rock gradually becomes calcareous, and at length a pure limestone. Then we have a mixture of argillaceous matter, which increases till a dark-colored clay, or mud, called shale, is produced. For a description of the fossils in the silurian and Devonian classes in our country, we are mainly indebted to the labors of Professor James Hall. — See his two splendid volumes on the Paleontology of New York.

The Potsdam sandstone is supposed to be the oldest fossiliferous rock in the United States, as well as Europe. The most common shell in the Potsdam sandstone is a lingula (Fig. 1); a genus which has survived all the revolutions of the earth, and is still found in the ocean. This rock contains also some polyparia, and one species of marine plant, the Scolithus linearis (2). It is said to be identified at the west in Michigan and Wisconsin.

In the calciferous sand-rock, fossils are both rare and obscure. The most common and characteristic are the Ophileta, an encrinite, Orthocera, Lingula,

Pleurotomaria, Scalites (3), Maclurea, Bellerophon, Orthis and Orbicula. It contains also marine plants, the Palæophycus tubularis (4). This formation extends south-westerly through the Atlantic states.

The Black River limestone embraces several varieties, varying from a dove-color to one almost black, as the Chazy, bird's-eye, Mohawk and Isle le Motte limestone, some of which varieties produce fine dark-colored marbles, as well as good quick-lime. One of its most characteristic fossils is the Maclurea, Columnaria sulcata (5), numerous Orthocerata (6), and a Fucoid, Phytopsis tubulosum (7), as well as many other kinds of fossils.

The Trenton limestone is a dark-colored, fine-grained and often shaly rock, very well characterized, and rich in organic remains. Here we first find that singular class of animals, the Trilobites, fully developed, such as Isotelus gigas (8), Calymene senaria (9), Trinucleus concentricus (10). Of shells we find a great variety, such as the Leptæna alternata (11), Orthis pectinella (12), Pterinea (Ambonychia) undata (13), Atrypæntous (14), Murchisonia bellicincta (15), Orthocera and Cyrtoceras annulatum (16).

The Utica slate is a black shale, the same as that interstratified with the Trenton limestone. It embraces a few peculiar fossils, as the Triarthus Beckii, a trilobite, and the Graptolites (17), which is generally regarded as a coral, and of which fourteen species are described in New York by Prof. Hall.

The Hudson River group consists mainly of shales and sandstones, formerly denominated Gray Wacke, not less than seven hundred feet thick. Some of its more characteristic fossils are Favistella stellata (18), Pterinea carinata, Cyrtolites ornatus (19), and Pentacrinites Hamptoni (20). Others are the Avicula demissa (21), Trinucleus caractaci (22), Strophomena (Leptæna) nasuta (23), and Modiolaris modiolis (24).

Mr. Logan, geologist of Canada, has lately described and figured numerous tracks of crustaceans on the Potsdam sandstone of that country. They exist also in the United States, according to Professor Hall. — (*See Quarterly Journal of Geological Society* for August, 1852.)

By looking at the appended geological map of North America, it will be seen that the most north-easterly spot where the lower silurian rocks are represented, is the Island of Anticosti, at the mouth of the St. Lawrence. Thence a strip extends along that river from its mouth to its source in Lake Ontario, and thence along the north side of that lake to Lake Huron. Another branch runs southerly, along the east side of the granitic mountains of Essex, in New York, along the borders of Lake Champlain and down the Hudson, and thence south-westerly to Alabama. These rocks, indeed, flank the mountains of Essex, so as to form a huge granitic island, giving us an idea of the state of things when the silurian rocks were in the course of deposition. Other patches of these rocks are shown in Pennsylvania, Ohio and Tennessee. In Missouri and Iowa, and extending north to the limits of the map, a vast region is colored as if lower silurian; as is done in one or two other spots; *e. g.*, in the vicinity of Hudson's Bay. But in all these cases the red color, with dots, comprehends whatever there may be of the upper silurian, the Devonian, and, with one exception, the carboniferous groups, and, indeed, all the other fossiliferous rocks. Nor can it be expected that the outlines of these fossiliferous strata are very accurately represented; and, doubtless, when the geology of the western part of our continent shall be minutely traced out, there will be found islands of non-fossiliferous rocks; as much variety, indeed, as the eastern part of the United States exhibits.

The Lake Superior sandstone, with its dykes or interstratified masses of trap, and which is so well known for its veins of copper, I have marked as lower silurian, following the opinion of Messrs. Foster and Whitney, sustained as it is by Professors Hall and Owen, and Mr. Logan, of the Canada survey. It is well known, however, that this opinion is disputed by some who have studied these rocks with care. — (*See the remarks of J. Marcou before the Geological Society of France, at its sitting, Dec.* 2, 1850.)

M. Boué has marked an extensive region in California and Oregon as silurian. But I do not find much evidence, in the reports of several military gentlemen and others on these regions, of the existence of rocks to any great extent older than the Devonian. Or, if they exist there, they have been so metamorphosed, as in the gold regions, that they come under my first class, or they are covered by cretaceous or tertiary strata. I have ventured, therefore, to strike out the silurian rocks from the Pacific slope, and extend the hypozoic and metamorphic strata to the tertiary, which we know is found of some width in California along the coast. I do not doubt that, when those vast regions shall be carefully explored by geologists, silurian strata will be found there, perhaps in great abundance; but I rather anticipate that they are much covered by tertiary deposits.

UPPER SILURIAN SYSTEM.

Following the authorities above quoted, the lowest member of the upper silurian in this country is the Oneida grit, five hundred feet thick; next, the Medina sandstone, three hundred and fifty feet; next, the Clinton group, eighty feet; next, the Niagara group, two hundred and sixty-four feet; next, the Onondago salt group, from six hundred to one thousand feet; next, the water-lime group, one hundred feet; next, the Pentamerus limestone, eighty feet; next, the Delthyris limestone, two hundred feet; and next, the upper Pentamerus limestone, which, according to Verneuil and Hall, corresponds to the upper part of the upper silurian strata in Europe.

Total thickness of the upper silurian, according to the above enumeration, about two thousand four hundred feet.

This arrangement embraces in the upper silurian the two highest members of the Champlain division, all the Ontario division, and more than half of the Helderberg series of the New York geologists, and all the Levant and most of the Premedidial series of the Professors Rogers.

The Oneida grit is coarse and fine grained, almost destitute of organic remains, save a small species of Terebratula and some Fucoides.

The Medina sandstone is a red or variegated siliceous mass, sometimes marly and friable, interstratified with gray bands of quartzose sandstone. In the red portions are numerous marine plants, and marine shells in the gray part. Of the plants, the most common and striking is the Fucoides Harlani (25). Of the shells, we find, Pleurotomaria pervetusta, Orbicula parmulata, Lingula cueaata (26), and Cypricardia orthonota.

The Clinton group is composed of red and variegated shales and sandstones, so diversified as to have received the name of Protean. It contains a most important bed of oolitic iron, which in Pennsylvania is worked to great advantage. The formation abounds in fossils, of which the Pentamerus oblongus (28), Delthyris brachyonota, Orthis circulus, Atrypa congesta (29), Nucula mactriformis (30), Actinocrinus plumosus (31), Fucoides bilobata, &c., may be named.

The Niagara group consists of shale below and limestone above. The former, exposed to atmospheric and aqueous action, crumbles away and leaves the limestone in overhanging masses, which at length break by their weight. These are the rocks over which the water at Niagara Falls is precipitated, and this is the cause of their retrocession.

This rock is highly fossiliferous, the shale abounding in shells, trilobites and crinoidea, and the limestone in corals. Among the trilobites are Asaphus limulurus (32), Calymene Niagarensis, Bumastis Barriensis and Homalonatus delphinocephalus. Among the shells are Strophomena subplana (33), Delthyris Niagarensis (34), and several other species. Among the encrinites are Caryocrinus ornatus (35), and Cyathocrinites pyriformis (36).

The Onondago salt group is an immense mass of argillo-calcareous shaly rocks, abounding in veins and beds of gypsum, and the source of all the salt springs in New York and the Western States. Notwithstanding its great thickness, however, it is very barren in fossils. But it abounds in hopper-shaped cavities, once probably occupied by crystals of rock-salt.

The remaining varieties of the upper silurian are placed by Professor Hall under the name of Lower Helderberg limestones.

The water-lime group derives its name from its

use for hydraulic cement, a portion of it, four or five feet thick, being useful for that purpose. The typical fossils of this group are Delthyris plicatus, Avicula rugosa (37), Tentaculites ornatus (38), Littorina antiqua, Atrypa sulcata, and Cytherina alta.

The Pentamerus limestone derives its name from a helmet-shaped pentamerus (the P. galeatus), (39), in which it abounds. In it occur also the Euomphalus profundus (40), and the Atrypa (Terebratula of Europe) lacunosa (41).

The Catskill or Delthyris shaly limestone consists of coarse uncrystalline limestone, and slaty as well as argillo-siliceous limestone. It is, perhaps, more fossiliferous than any rock in the country of equal thickness, abounding in species of Delthyris, Atrypa, Acidaspis, Asaphus, Favosites, Calceola Conularia, &c.

The upper Pentamerus limestone is considered distinct from the Delthyris shaly limestone by its organic remains. It contains a smooth Pentamerus, also several forms of Atrypa.

A reference to the accompanying map of North America will show that the upper silurian rocks are largely developed in New Brunswick and Lower Canada. They form also a part of the surface in Cape Breton and Nova Scotia, according to the map of those countries by Dr. Gesner. —(*Quarterly Journal of the Geological Society*, vol. I., p. 23.) Passing south-westerly, we first meet with them in New York, from whence a strip is seen running westerly, along the south side of Lake Ontario and the north side of Lake Erie, whence it spreads out southerly and westerly over more than half of the Western States; and when the country towards the Rocky Mountains has been thoroughly examined, no doubt the upper silurian rocks will be found largely developed there, and perhaps on the west side of the mountains, as already remarked. A band of these rocks, as will be seen, follows the Appalachian chain of mountains from New York to Virginia, and, indeed, in a strip too narrow to be represented on the map, to Alabama.

DEVONIAN SYSTEM.

The lowest member of the Devonian system, according to Verneuil and Professor Hall, is the Oriskany sandstone, seven hundred feet thick, in Pennsylvania; the next is the Cauda Galli grit; next, the Schoharie sandstone; next, the Onondago limestone, fourteen feet thick; next, the Corniferous limestone, seventy feet thick; next, the Marcellus shale, fifty feet thick; next, the Hamilton group, one thousand feet thick; next, the Tully limestone, twenty feet thick; next, the Genesee slate, two hundred and fifty feet thick; next, the Portage group, one thousand feet thick; next, the Chemung group, one thousand five hundred feet thick; next, the old red sandstone, perhaps three thousand feet thick. Total thickness, about seven thousand six hundred feet.

In lithological characters, the Oriskany sandstone is usually a tolerably pure siliceous sandstone, of a white or yellow color, passing, however, into a tough siliceous limestone. Its most common fossils are the Delthyris arenosa (42), and Atrypa elongata (43).

The Cauda Galli grit derives its name from a peculiar fucoid, which it contains, resembling the tail of the domestic cock, or rooster. The Schoharie grit is a calcareous sandstone, abounding in a species of Pleurorhyncus and Orthocera, with many corals.

The Onondago limestone is a tolerably pure limestone, forming an elegant marble, and its organic remains are very abundant and widely dispersed, as is the formation. A large part of the rock is frequently made up of fragments of crinoidea and corals. Among these are Favosites alveolaris (44), and the large smooth Entriochite, called Encrinites lævis (45); also, Favosites fibrosa (46), Astrea ragosa (47), and a Cyathophyllum (48). In this rock are found the defensive fin-bone of fishes, which, with one exception, in the case of the Oriskany sandstone, is the lowest place in the rocks of our country where vertebral animals have been found.

The Corniferous limestone derives its name from the number of flinty and hornstone nodules which it contains. These are usually more or less arranged in layers. Instead of the corals of the last-named rock, we find here little else but shells. Along with the Atrypa, Strophemena and Delthyris, we find some that are rather peculiar, as the Paracyclas elliptica, Pleurorhyncus trigonalis (49), Pterinea cardiformis (50), Orthonychea, a new genus (51), Euomphalus (?) rotundus, Cyrtoceras undulatus (52), Strophomena undulata (53), &c.

Professor Hall calls the two last-named rocks Upper Helderberg limestones; the four next he denominates the Hamilton group.

The Marcellus slate is black and bituminous, sometimes calcareous, and so much like the shale of the coal formation as often to excite expectations of finding that mineral, not to be realized. Its most characteristic fossils, which are not numerous, are the Orthoceras subulatum (54), Strophomena setigera, &c., Avicula muricata (55), and Goneatites expansus (56), Orthis, Orbicula, Tentaculites, &c.

Dull olive, bluish-gray, calcareous shale, constitutes the principal mass of the Hamilton group. It contains remarkable septaria, called turtle-stones, from their great resemblance to petrified turtles. The nucleus is either a fossil body, or iron pyrites, and they sometimes attain the size of one or two feet. Its organic remains are numerous, such as Dipleura, a trilobite, Orthonota undulata, Delthyris mucronata (57), Orthocera constrictum, Avicula flabella (58), Tellina (?) ovata (59), Turbo lineatus (60), several Nuculæ, the Cucullea Modiola, Cypricardia bufo (61), Delthyris medialis (62), Strombodes helianthoides (63), &c., &c.

An impure, dark-colored, thick-bedded limestone,

called the Tully limestone, succeeds to the shales of the Hamilton group. Its fossils are not abundant. The Atrypa cuboides and Orthis resupinata, however, are peculiar to this rock.

The Genesee slate, or group, is little else but indurated mud, charged with bitumen. It attains a thickness sometimes of two hundred and fifty feet, and is very liable to disintegration. Its organic remains are Orbicula, Lingula, Avicula, Strophomena, Tentaculites, and a linear grass-like plant, probably a sea-weed. It is the Postmedidial newer black slate of the Rogerses.

The Portage or Ñunda group (the Postmedidial Flags of the Rogerses) consists of black shales below, and sandstones above. Its thickness is estimated at one thousand feet. It abounds in concretions, some of which (as the one called "cone in cone," *Hall's Report*, p. 232) have not yet been well explained, as to whether they are mere concretions or petrifactions. Here also are found those large concretions called turtle-stones, sometimes several feet across. The erosions in this rock by streams produce some of the most striking water-falls and gorges in our country.

Organic remains are not very abundant in this group. The Fucoides graphica is perhaps the most characteristic. Among the shells we have Aviculas, Ungulina suborbicularis (64), Bellerophon, Orthoceras, Goniatites, Pinnoposis acutirostra (65), Delthyris, Orthis, Nucula, and the beautiful Cyathocrinus ornatissimus.

The Chemung group consists of a series of thin bedded sandstones, with interstratified shales and impure limestones. The greatest thickness is fifteen hundred feet. Its organic remains are somewhat abundant. Among these we find one trilobite, the Calymene nupera, several beautiful species of Avicula, Pecten, Lima, Strophomena, Orthis, several species of Delthyris, Atrypa, &c. Several species of vegetables have also been found, such as Sphenopteris and Sigillaria, which are land plants.

In Europe the Devonian group is usually regarded as only "various modifications of the old red sandstone series" (De la Beche). But Professor Hall and other geologists in our country speak of the latter rock as only one of the varieties of the former. In this restricted sense it is largely developed on this continent. In New York, the Catskill Mountains, as much as three thousand feet high, are mainly composed of it. It consists of various strata of sandstone, shale and shaly sandstone; the sandstones being of a red or reddish color.

The organic remains of this sandstone are quite different from those in the rocks below. The most characteristic are the scales of certain fish, particularly the Holoptichius nobilissimus, found abundantly in the old red sandstone of Great Britain and Russia.

A reference to the map shows that the Devonian system has a wide development in our country. We see a patch of it at the very eastern extremity of Lower Canada, on the south side of the St. Lawrence. Next we see it largely developed in New York and Pennsylvania. From thence two branches extend south-westerly, on each side of the great coal-field of the Appalachian Mountains, and reaching even into Tennessee and Alabama, though I have attempted to represent it only on the western side of the coal. In Michigan, also, it is enormously developed; and it surrounds a large part of the coal-field of Illinois, as well as that of Missouri and Iowa.

CARBONIFEROUS SYSTEM.

The two most important members of this most valuable of the rock formations are the carboniferous limestone, forming the basis, and the coal measures above. The latter, however, has at its lower part a formation usually considered distinct, called the millstone grit; and beneath this, in some places, we have a group of red shales and shaly sandstones.

CARBONIFEROUS OR MOUNTAIN LIMESTONE.

Although in Europe this rock is simply a limestone, alternating in its upper part sometimes with coal; yet in this country, especially if we include in it the gypsiferous formation of Nova Scotia, limestone is its least important member, its place being taken by sandstones of various colors, and by gypsum. Its true character, however, is determined by its organic remains, and by its position, which is immediately below the coal measures, as in Europe. In Nova Scotia it contains some trilobites, such as a limulus, the Nautilus, Orthoceras, Conularia, Euomphalus, Cypricardia, Isocardia, Cucullæa, Avicula, Pecten, Terebratula, Spirifer, Producta, Encrinus, Cyathophyllum, Favosites, &c.

An inspection of the map of the United States will show that this formation is developed in narrow belts around all the important coal-fields of this country, including Nova Scotia and New Brunswick. In Europe the chief deposits of lead occur in this rock. But Professor Owen does not admit that it is the lead-bearing rock of this country. — (*See his Report of a Geological Exploration of part of Iowa, Wisconsin and Illinois*, p. 19). He regards the rock of those states yielding lead, copper, &c., as a limestone, called the Upper Magnesian Limestone, peculiar to the Western States, and coming in between the Trenton limestone and the Utica slate of the New York survey, and a part of the Lower Silurian system. — (*See his final Report on Wisconsin, Iowa and Minesota*, 2 vols., quarto, 1852.)

The researches of Captain Stansbury around the Salt Lake of Utah, according to the examination of the specimens by Professor Hall, prove the existence of carboniferous limestone there extensively, and I have accordingly marked a spot around that lake as of this rock ; but without any means of knowing its extent, or of representing the metamorphic rocks with which it is associated.

COAL MEASURES.

The great coal formation of this country consists, like those in other parts of the world, of beds of coal, from a mere line up to fifty feet in thickness, interstratified with sandstone, shale and limestone, which constitute the great body of the formation. It is enormously developed in North America, as an inspection of the geological map will show. We have first, in the north-eastern part, in Nova Scotia and New Brunswick, a coal-field covering not far from ten thousand square miles. In the south-east part of Massachusetts, and in Rhode Island, is a deposit covering about five hundred square miles, which, we have become convinced, and, were there room, we think we could easily show, is a metamorphic coal-field, and, although as yet but little explored, we cannot doubt but it will hereafter be found to contain important seams of workable coal. The great Appalachian coal-field, extending from New York to Alabama, seven hundred and twenty miles in length, covers nearly one hundred thousand square miles. The Indiana coal-field, three hundred and fifty miles long, embraces about fifty-five thousand square miles. Another occurs in Michigan, one hundred and fifty miles long, covering about twelve thousand square miles. The Missouri and Iowa coal-field, which has been mapped by Professor Owen in his Report, and which I have transferred to the map accompanying this work, may be set down at fifty thousand square miles. The grand total, to say nothing of fields yet further west, amounts to more than two hundred and twenty-five thousand square miles, equal to about twenty-eight such states as Massachusetts! If we suppose the average thickness of all the beds to be fifty feet (some single beds are as thick as that), the whole amount, in solid measure, of the coal in the United States, would be THREE MILLIONS AND A HALF of cubic miles! — a quantity absolutely inconceivable, yet the calculation is certainly a moderate one.

Whatever else, therefore, fails in the United States, her coal can never be exhausted; and when we think of the immense extent to which the use of steam will hereafter be increased on this continent, in consequence of the use of coal, we can form no adequate conception of the future populousness and prosperity of this country. If we suppose our numbers to become as many only to the square mile as now are found in Great Britain, namely, two hundred and fifty, the United States, east of the Rocky Mountains, would contain nearly five hundred millions, and, including Oregon and California, more than seven hundred millions, equal almost to the present population of the globe.

In view of such statements, it is no wonder that almost every community feels a deep interest in the discovery of beds of coal. And the fossils found in the coal measures, when not obscured or destroyed by metamorphic action, enable the geologist to decide with great confidence whether a series of rocks belongs to the true coal formation. These fossils consist of certain well-marked and peculiar land plants. Those most characteristic are ferns (66), Sigillaria (67), which are probably for the most part the trunks of large tropical ferns; stigmariæ (68), which are the roots of the Sigillaria; Lepidodendra, allied to existing club mosses or ground pines (69), Calamites (70), which are allied to the Equisetaceæ of the living flora; Coniferæ and Cycadeæ, and asterophyllites (71), whose true place among plants is not fully settled.

Captain R. B. Marcy, of the United States army, has recently (November 1852) put into my hands the specimens he collected in his expedition last summer to the sources of the Red river, with an outline map. On that map, at Fort Belknap, on Brazos river, I find it stated that "large beds of bituminous coal" occur. From that statement I was led to examine the reports of other explorers of our western country, and I have found four other places, lying to the north and north-west, where coal occurs. The extreme points are nearly eight hundred miles apart, and the localities, with one exception, are not far from the eastern foot of the Rocky Mountains. Although these deposits do not all probably belong to the same age, I cannot but suspect strongly that several of them may be outcrops of a genuine and very extensive coal formation, which is hidden perhaps the greater part of the distance by newer formations. If this suggestion should prove true, it may be of immense importance to the future prosperity of that portion of our country; and I have indicated the spots where the coal occurs upon the accompanying map of the United States, as follows:

In 1835, Rev. Samuel Parker described "anthracite coal" as occurring on the north branch of Platte river, first considerably east of the Black Hills, say in latitude one hundred and four degrees west, and north latitude forty-two degrees; and then repeatedly, for several degrees, say nearly to the Red Buttes, in longitude one hundred and seven degrees, and latitude forty-three degrees. He says it was "the same, to all appearances, as he had seen in the coal-basins of Pennsylvania," and he found it in beds.

Colonel Frémont found coal and fossil plants on the same route, in west longitude one hundred and eleven degrees, and in north latitude forty-one and a half degrees. Eleven species of these, which are ferns, are figured by Professor Hall, and many of them greatly resemble those in the true coal measures. He also found coal (brown coal, probably) in longitude one hundred and seven degrees, latitude forty-one and a half degrees.

Major Emory met with "bituminous coal in abundance" in west longitude nearly one hundred and five degrees, and north latitude forty-one degrees, on the Arkansas. He was told of one bed thirty feet thick.

Lt. Abert found shale, which he regarded "indubitable proof of the existence of coal," in west longitude one hundred and four and a half degrees, and north latitude thirty-six and a quarter degrees.

In 1818, Mr. Bringier described "a large body of blind coal (anthracite), equal in quality to the Kilkenny coal, and by far the best he had seen in the United States, immediately on the bank of the Arkansas, a little above the pine bayou, five hundred miles from its mouth." — (*Am. Journal of Science*, vol. III., p. 41.)

In the north-west part of Texas, Dr. F. Rœmer describes an immense deposit of cretaceous rocks, which surround, between the Pedernales and San Salo rivers, granitic and palæozoic formations, the latter of carboniferous limestone and silurian rocks. — (*Am. Journal of Science*, vol. VI., p. 26. New series.)

These statements then present us with two localities of anthracite and two of bituminous coal, which it would seem persons not skilled in mineralogy could hardly fail to distinguish. Both the cases mentioned by Frémont were probably brown coal, of the newer strata. But, setting these aside, I still think there may be some plausibility in the suggestions I have made.

TRIAS, OR NEW RED SANDSTONE.

These rocks in Europe are now usually separated into two divisions considered as quite distinct; namely, the Triassic system, or upper new red sandstone, and the Permian system, or lower new red sandstone : the last belonging to the Palæozoic rocks, and the first to the Mesozoic group. In this country we have not yet been able to recognize such a division. But we have the new red sandstone somewhat widely diffused, and some of it corresponds with the Triassic series of Europe. It consists mainly of red, gray and white sandstones, and conglomerates and shales, with limestone occasionally interstratified. The most northern of these deposits occupies the valley of Connecticut river, across Massachusetts and Connecticut. Another deposit commences in New Jersey, and extends, with some interruptions, into the southern part of Virginia, as may be seen upon the map.

The evidence that this series of rocks belongs to the new red sandstone is, first, *its lithological characters*. These correspond, not, indeed, exactly in all cases, but yet generally, with the new red sandstone of Europe, as I have largely shown in my *Final Report on the Geology of Massachusetts*, vol. II., p. 435. Secondly, *its organic remains*. The most important of these are fishes, which have heterocercal tails (that is, the upper lobe longest), and therefore must have come from a rock older than the oolite, according to the views of Agassiz. Thirdly, *the relative position of this rock*. In New England, indeed, it rests upon the older crystalline rocks; but in New Jersey, Pennsylvania and Virginia, it is found to lie above the coal measures. Now, as the character of the fishes places it below the oolite, we have great reason to conclude it to be new red sandstone; and most likely we have both the Trias and the Permian systems in these deposits.

The tracks of animals found in this rock are truly remarkable. Excepting in a single instance, in New Jersey, they have not been discovered out of the valley of Connecticut river. But there not less than fifty species have been recognized, varying in size from half an inch to twenty inches in length. The greater part of them were made by bipeds, most of them, probably, birds. But at least a dozen quadrupeds can be recognized, most of them with hind feet much larger than the fore feet. Figs. 72 to 83 will give an idea of the outlines of several species of these tracks. But, for a full description, with figures of the natural size, of all the species, I would refer to a paper in volume III., new series, of the *Transactions of the American Academy of Arts and Sciences*, where I have attempted to name and describe the animals that made the tracks. Figs. 72, 73, 74, 75, 76, 81, 82 and 83, were made by bipeds. Figs, 77, 78, 79, 80 and 84, by quadrupeds.

Connected with this deposit of sandstone, almost everywhere along the Atlantic coast, we find ranges of trap rock, mostly of that variety called greenstone. It seems to have been protruded to a considerable extent after the deposition of most of the sandstone, and along curved lines, though very rarely cutting directly across the strata. In the vicinity of the trap the sandstone has often been indurated, rendered vesicular, and otherwise metamorphosed. In the Connecticut valley we find, also, a rock called *volcanic tufa*, composed of fragments of the trap and the sandstone, yet evidently of volcanic origin. But it contains vegetable stems converted into vesicular trap, yet retaining their form.

DEPOSIT SUPPOSED TO BE OF THE AGE OF THE OOLITE.

Many of the limestones of this country have an oolitic structure, — that is, are made up of minute spherical concretions, — but they are not, therefore, identical with the genuine oolite of the eastern world, for they have not the same position on the geological scale, nor similar organic remains. But in eastern Virginia is a coal-field many miles in extent, with coal seams sometimes thirty feet thick; and the evidence seems quite decided that these sandstones and shales are of the age of the oolite, or rather of Lias, regarding that as a variety of oolite. No two rocks can be more unlike in lithological character than these coal measures and the lias of of the eastern continent. But the plants and the fishes which the former contain correspond essentially with those of the latter; and animal and vegetable types are a much surer guide, as to the age of a rock, than its mineral composition. As to superposition, the rock of Virginia reposes on granite and crystalline slates, and is overlaid only by alluvial deposits, or perhaps drift.

CRETACEOUS SYSTEM.

On the eastern continent the most obvious peculiarity of this formation, from which its name is derived, is the occurrence of chalk in it, although other varieties of rock are not wanting. But we have no evidence of any genuine chalk existing naturally in this country. We have, however, strata immensely developed, whose organic remains correspond, at least generically, with those of the chalk, and hence they must be referred to that division on the geological scale. New Jersey is the most northerly point on this continent where these rocks occur, and there they consist of green and ferruginous sands, with only here and there an overlying deposit of marly limestone. As we go south, however, the formation expands and the limestone predominates, while it is only here and there that the subjacent sandstone is visible. As we advance into Texas, the limestone becomes hard and compact. In lithological as well as organic characters, this formation is said to correspond, to a considerable extent, with the same formation on the eastern continent; the strata, for instance, of Texas, resembling the cretaceous deposits around the Mediterranean, while those of the Atlantic coast agree more nearly with those of England and northern Germany.

Dr. Morton divides the cretaceous strata of the United States into three groups; the upper, medial and lower. The characteristic fossils of the upper division are Nummulites Mantelli (84), Terebratula lachryma, Pecten membranosus (85), P. Poulsoni, Plagiostoma dumosum, Balanus peregrinus, Nautilus Alabamensis, Scutella Rogersi, S. Lyelli and Echinus infulatus. Those of the medial division are Spatangus parastatus (86), Nucleolites crucifer, Ananchytes fimbriatus, Teredo tibialis, Scalaria annulata, Belemnites ambiguus and vermetus rotula (87). Also, remains of the huge Zeuglodon cetoides, a cetaceous animal, sometimes one hundred feet long. Remains of crocodiles and sharks also occur here, as teeth of the Lamna plicata (88), and L. obliqua (89), and of a crocodile (90). Those of the lower division are the Belemnites Americanus (91), Exogyra costata, Gryphæa convexa (92), Ostrea falcata (93), Ammonites Delawarensis, Hamulus onyx (94), with remains of saurians and fishes.

In the cretaceous deposits of Texas Dr. Rœmer finds three species of Hippurites, which is entirely wanting in the formation in New Jersey, but abundant around the Mediterranean, as on Mount Lebanon; also the genus Caprina, which is wanting in New Jersey and Alabama, but abundant and characteristic in Texas (*Am. Jour. of Science for July*, 1848, p. 25), and also around the Mediterranean.

It is not possible, at present, to mark with accuracy on the map the limits of the cretaceous formation in the United States. On the Atlantic coast, commencing in New Jersey, it has been traced out accurately; but in the southern and western states we have only now and then an observation of some geological traveller to guide us. In Texas I have followed Dr. Rœmer, as far as could be done from his descriptions, unaccompanied by a map. Still further north, the specimens collected by Colonel Frémont, at the foot of the Rocky Mountains, in latitude thirty-eight degrees, and longitude one hundred and five degrees, and, indeed, from ninety-six degrees to one hundred and five degrees, show the whole distance to be a cretaceous deposit. On the north, I have depended upon the researches of M. Nicollet, who found cretaceous fossils a thousand miles beyond Council Bluffs, at the mouth of the Yellowstone river. I know of no reason why the same formation may not extend much further to the north, along the eastern base of the Rocky Mountains. Doubtless older formations will be found to protrude occasionally through the cretaceous, as we know to be the case from the statements of Dr. Rœmer in Texas. But it will require much time and labor to trace out all these details, and for the present we must be contented with such a general and doubtless quite imperfect view as the annexed map exhibits. We may be quite sure that there is an immense development of this formation in our country.

TERTIARY STRATA.

Those geologists who have examined our tertiary strata with most care are of opinion that they may be divided, as they have been in Europe, into the Eocene, the Miocene and newer Pliocene. These distinctions, however, founded as they are upon characters quite arbitrary and indefinite, will not probably be permanently retained, and will be supplanted by others more natural. Some arrangement of this sort is convenient to indicate the different ages of these deposits, for a long period must have been occupied in their production.

Of the groups of these strata, the two lowest, the Eocene and Miocene, have been most studied ; yet, as they are shown on the accompanying map, probably only a small part of the surface actually underlaid by them is exhibited. I have followed the delineations made by Mr. Lyell, in his travels in the United States. No attempt is made to mark out the Pliocene, with a single exception.

All the Post Pliocene and tertiary deposits have but one color on the map ; and wherever the Miocene and Eocene tertiary are known to exist, they are designated, the first by dots and the last by crosses. With this exception, I have made no attempt to distinguish between the tertiary and alluvial.

Miocene Strata. — The most northerly point at which these strata occur in the United States is upon the Island of Martha's Vineyard. They were first described by me in 1823 (*Am. Jour. Sci.*, vol. VII., p. 240), and referred to the plastic clay formation of Europe, which is Eocene ; but Mr. Lyell thinks they are probably Miocene (*Am. Jour. Sci.*, vol. XLVI., p. 318). South of this place we find these strata extending from Delaware Bay to Cape Fear, occupying portions of Delaware, Maryland, Virginia, and North Carolina, embracing an area of about four hundred miles long, and from ten to seventy miles broad. Patches of the same formation appear in South Carolina and Georgia.

These strata consist of sands and clays of various colors, often very striking, as at the cliffs of Gay Head, on Martha's Vineyard. On James river, at Richmond, Virginia, is an interesting deposit of infusorial earth, ten to twenty feet thick, and extending over an unknown area. The proportion of fossil shells identical with those now living is said to be about seventeen per cent. The genera of shells and fossils generally correspond with those of the Miocene strata of Europe, but the species very rarely. Remains of the whale tribe are not uncommon, and those of the mastodon longirostis have been found.

Eocene Strata. — These consist of limestones and marls in their lower part, and of burrstone, with red loam, mottled clays, and yellow sand, in their upper part. The most northern point where they have been described is in Maryland, on the Potomac, from whence they extend southerly through Virginia, South Carolina and Georgia. The sterile sands forming the pine barrens of Virginia and North Carolina appear to belong to the Miocene strata, while similar barrens in South Carolina and Georgia are Eocene.

Of the species of Eocene shells found in the United

States, not more than five per cent. have been identified with those described in Europe, and a still smaller proportion correspond with living species.

Most of the low and level lands on the Atlantic coast of the United States are composed either of tertiary or cretaceous rocks. Next succeeds a more elevated and less level region, which is hypozoic or crystalline. Of course, the rivers which cut across all these formations usually show rapids or cataracts where they pass from the crystalline to the newer deposits. Hence such places usually form the head of navigation for vessels, and it is this circumstance that has located so many large cities on the line between the hypozoic and the recent rocks; as New York, Trenton, Philadelphia, Washington, Richmond, Augusta, Columbus, Wetumpka, &c.

It has long been known to geologists that numerous deposits of brown hematite iron ore, associated with ochres and mottled clays, occur, in connection with a highly ferruginous limestone and micaceous and argillaceous slates, through the whole distance from Canada to Georgia; lying along the west side of the crystalline and hypozoic strata of the Green, Hoosac and some ranges of the Appalachian Mountains. No geologist has doubted that these deposits were all contemporaneous; but their true age has been a mystery. During the year 1852, my attention was drawn to one of these deposits, in Brandon, Vermont, which has the peculiarity of containing a bed of carbonaceous matter twenty feet thick, with fifteen or twenty species of fossil fruits. The leading result of my examination is, that the Brandon deposit belongs to a tertiary formation, and the carbonaceous matter is very much like the brown coal of Germany. And since that, for the most part, belongs to the Pliocene or newer tertiary, we may provisionally place the Vermont deposit in the same place, and, by parity of reasoning, all the brown hematite beds, extending at least twelve hundred miles through the United States. So confident am I of the soundness of these conclusions, that I have ventured to mark a strip of tertiary on the map near the line along which the hematite beds occur, although they are not always in a continuous line, but scattered over a considerable breadth of surface. I do not, however, regard it so certain that this deposit is Pliocene tertiary, that I have ventured to mark it as such, but only as tertiary.

The fruits of the Brandon deposit are beautifully preserved, but they are quite peculiar, and as yet their affinities have not been made out. A few of them are represented on the plates (95, 96). A full account of them, with the inferences, is given in the *American Journal of Science* for January, 1853.

DRIFT.

North America corresponds to Scandinavia in the immense development of drift, and, since the underlying rocks are essentially alike in the two regions, the minutest features of the phenomena correspond.

In Scandinavia, however, there appears to have been a centre of dispersion, so that the lines of drift radiate. But, if such a centre exist in North America, I do not think it has yet been discovered, although some have supposed the White Mountains to have been such a centre. That glaciers may have existed there, when the climate was colder, and at the drift period, is probable. Yet I have found that the striæ, at least five thousand feet high on those mountains, have a north and south direction, as in other parts of New England.

During the year 1852, however, I have discovered several examples in Massachusetts where Hoosac Mountain slopes into the valley of the Connecticut, in which very distinct traces of ancient glaciers exist. Whether they occur on the west side of that mountain, I have had no opportunity of ascertaining; but my present impression is that the Green and Hoosac Mountains once constituted a crest, from which glaciers originated. A description of these cases I have just given in a Report on Geology to the government of Massachusetts.

From the eastern extremity of Nova Scotia to the Rocky Mountains, over a belt several hundred miles wide, the surface is strewed over with boulders, gravel and sand, that have been carried in a southerly direction by the drift agency, from a few rods up to five hundred or six hundred miles. The rocks in place, also, have been striated in the same direction. This course is usually a few degrees east of south and west of north, but sometimes as much east of north and west of south.

The distance southerly to which this drift agency extended has not been accurately fixed. Indeed, it seems to have gradually died out, and in some places it extended much further than others. South of the Ohio river not much coarse drift appears, and yet, much further south, in the valley of the Mississippi, pebbles occur which had a northern origin. Perhaps the latitude of forty degrees north is the southern limit of a decided drift agency; and accordingly the lines that represent that agency on the geological map are terminated, for the most part, on that parallel.

As to the height above the present ocean to which the drift agency extended, we have only a single example from which to judge. But that is a decided one. Mount Washington, the highest point of the White Mountains, about six thousand two hundred feet, does not show distinct marks of this agency much over five thousand feet elevation. Above this the surface is covered by angular blocks, broken up by frost, but never removed; a circumstance that rarely, if ever, occurs at a lower level. Between five thousand and six thousand feet, then, may be assumed as the upper limit of this agency in North America.

No organic remains have been found in the proper drift of this country. Recent species of shells have, indeed, been discovered in sand and gravel, which have been called drift; as at Brooklyn, Long Island, and at Portland, Maine. But these deposits were probably reärranged and modified drift, and were formed at a period immensely posterior to that of the proper drift.

ALLUVIUM OR POST PLIOCENE FORMATION.

This formation, as its name implies, embraces all the deposits that are newer than the tertiary, though I have described drift as distinct. Recent examinations and measurements of these deposits in New England have greatly modified my own views of their origin, and of the changes which the surface of this country has undergone since the drift period. I can give, however, in this place, only a brief summary of the conclusions at which I have arrived.

The most genuine examples of drift occur upon our mountains. It consists of blocks of stones, usually but little rounded, with coarse sand and gravel, mixed promiscuously together, but rarely sorted and stratified. It is, in fact, the same materials as glaciers tear off from the rocks over which they slowly move, and which they crowd along as moraines. And, in fact, the drift materials were probably abraded and pushed along either by glaciers or enormous icebergs, that acted as glaciers, — most likely both were concerned. As we descend from the higher parts of the mountains of New England, — and the same thing is true in Europe, — we find that aqueous and glacial agencies, mainly the first, have acted on the drift, grinding it into finer materials and sorting the coarser and finer parts, and spreading them along the sides of the hills in the form of beaches, more or less level-topped. Descending still lower, this process seems to have gone on more thoroughly, and we come, at length, to genuine level-topped terraces, — sometimes of thoroughly rounded gravel, sometimes of coarse or fine sand, then of clay, and, lowest of all, of loam, such as constitutes alluvial meadows. The lower we descend, the more completely has the process been carried on, and we see it yet going on along our rivers. At all heights, however, we shall find many places where the same kind of drift occurs as on the tops of the hills; almost everywhere, indeed, where the beaches and the terraces have been swept away, or have never been formed. And furthermore, scattered over and among the coarser terraces and beaches we frequently find large angular travelled blocks, that seem to have been transported by an agency somewhat different from that which sorted and comminuted the terraces and the beaches. But in all cases we find the terraces and the beaches above the great mass of the drift; and it cannot be doubted that the former have resulted from the reärrangement and comminution of the latter.

Now, how has this been done? It seems to us that no one can examine these phenomena in New England, — and Mr. Chambers' recent work on Ancient Sea Margin shows the same in respect to

Great Britain, — without being satisfied that water has been the main agent. For what other agent can sort and arrange with a level top, and in horizontal layers, gravel, sand and clay, over wide surfaces ?

But, were these waters the ocean, ancient lakes, or rivers ? Doubtless the phenomena require all three of these phases of aqueous action.

If I mistake not, I have already ascertained the existence of some of these ancient sea-beaches in the Hoosac range of mountains of Massachusetts, not far from two thousand feet above the present ocean, and twenty-five hundred feet in the White Mountains. Also, some river terraces upon mountain streams, as much as one thousand eight hundred feet above the ocean. Around some of our lakes (Ontario, for instance) we find distinct terraces as high as nine hundred and ninety-six feet above the sea, and seven hundred and sixty-four feet above the lake. As we descend from these highest points yet ascertained, we find beaches and terraces at various levels, till we reach the present sea level. Along the rivers we do, indeed, find numerous basins, whose outlets may have been blocked up ; and, as they were gradually worn away, the terraces may have been produced by the rivers themselves in their ordinary mode of action, in a climate where occasional floods occur. But, when the waters stood high enough to form the uppermost of the beaches and terraces above mentioned, — or even half or a quarter as high, — they would, in New England, have overtopped all the solid barriers that could have existed, and have communicated with the ocean. It must have been the ocean, therefore, by which they were produced; and the inevitable conclusion is, that the ocean, at the close of the drift period, did stand over this continent, as high as we now find any terraces or beaches; and that, as it retired, or the continent was lifted up, the terraces and beaches were formed, excepting some of the river terraces, which may have been produced as above described. The terraces and beaches around our lakes, and on our mountains, may have been the result of paroxysmal elevations of the surface; but along our rivers the terraces may have been produced by a slow, continuous erosion, in connection with freshets; as I have endeavored to show, in my Reports on the Geology of Massachusetts; but the reasoning would occupy too much space to be introduced here.

These views almost annihilate drift as a distinct formation; for most of the surface of our country is covered with drift that has been modified by long-continued aqueous action, though, doubtless, ice, both as glaciers and icebergs, was concerned, especially in the earlier stages of the process. The immense amount of these modified and sorted materials spread over the surface, and the great height to which we can trace the modifying process, show us how very far back we must place the drift period, and how very long has been the post-pliocene. For, if we do not carry back the latter to the time when the process began, we can draw no line between that and the previous formations.

If the United States were covered with the ocean so deeply at the drift period, and if similar facts exist in Europe, there seems to be no small reason to believe, as some geologists have attempted to show, that the ocean has retired from the land, although the land may also have been lifted up. Yet, if elevated from twelve hundred to two thousand feet, we might expect that some of the beaches and terraces would exhibit traces of the movement. But they show none. If the continent has been lifted at all so recently, it must have risen bodily, and very equably. Why not admit that refrigeration of the crust beneath the oceans may have deepened their beds and drained the land?

Another proof of the submergence of at least a portion of North America during a part of the post-pliocene period is the occurrence of marine shells in the strata deposited at that time. These have been found in the clays along the eastern side of Lake Champlain, and in the valley of the St. Lawrence river. This deposit extends around Champlain to the height of three hundred feet above the lake, which is about one hundred feet above tide-water. Only a very few species, however, and those of Arctic character, occur, and in a few places; and the upper part of the deposit is destitute of them. But their existence proves either that the region was beneath the ocean at that time, or that a great inland sea existed, whose waters were salt or brackish, and which became fresh as the waters were drained off by the cutting away of the barriers.

Another interesting process that has been going on during the post-pliocene period, though it commenced vastly earlier, is the denudation of the strata by the ocean, and the erosion of solid rocks by the agency of rivers. This work, indeed, commenced as soon as the rocks were consolidated, and thrown into essentially their present forms. Hence it runs back immensely.

The ocean, acting by its waves, as it has repeatedly risen over the surface and fallen back, has done most towards this work. It has been calculated, from accurate admeasurements, that the strata in South Wales, England, have, by this agency, been reduced as much as ten thousand feet. No doubt the surface of New England, and, generally, the mountainous parts of our whole country, have been worn away as much; and the present forms of most of our hills have mainly resulted from this agency. Sometimes the sea has worn out narrow gorges, called Purgatories, as at Sutton, Massachusetts, and Newport, Rhode Island. More often it has formed precipitous ledges, or widened the upper part of valleys. In this way, probably, the sandstone in the Connecticut valley has been worn back ninety miles, from New Haven to Mettawampe (Toby), and Sugar Loaf, at least twelve hundred feet deep.

There is evidence, however, that rivers have, in

some places, cut out deep channels in the rocks, all over the country. Such are the *canons* on our western rivers, sometimes two hundred and fifty feet deep and several miles long. Such is the cut, seven miles long, and one hundred and fifty feet deep, from Niagara Falls to Lake Ontario; and similar cuts on Genesee river, not only at its mouth, seven miles back, but from Mount Morris to Portage, fifteen miles, sometimes four hundred feet deep. Such are the Gates of the Rocky Mountains, near the source of the Missouri, six miles long and twelve hundred feet deep. Numerous cuts exist on the Columbia river, between the American Falls and the Dalles, — most of the way between these points being many hundred feet high. On the Connecticut river, at Bellows Falls and Brattleboro, there is evidence that the channel has been eroded six hundred or eight hundred feet, mainly by its own waters. To these might be added a multitude of examples from every mountainous part of the country, but space will not permit.

A third process of interest, that has been going on during the post-pliocene period, is the accumulation of alluvial matter at the mouths of our rivers, called deltas. By far the most remarkable one on this continent is that at the mouth of the Mississippi. This is several hundred miles long, and has advanced several leagues into the Gulf of Mexico since the settlement of New Orleans. It contains two thousand seven hundred and twenty cubic miles of matter; and, as now increasing, it must have required fourteen thousand two hundred and four years for its formation.

It seems to be now generally admitted by geologists that the bones of large extinct quadrupeds, which have been found in the superficial deposits of North America, occur in those of the post-pliocene period, although formerly referred to the drift. In Kentucky, South Carolina, New Jersey, Maryland, Mississippi, New York and Vermont, the Elephas primigenius has been found. In New York, Kentucky, Virginia, New Jersey, Connecticut, Mississippi, Ohio, Indiana, Missouri, and, indeed, in most of the Western States, occurs the Mastodon maximus. On Skiddaway Island, coast of Georgia, the Megatherium has been found; in Virginia and Kentucky, the Megalonyx; in Kentucky and Mississippi, several species of ox; in Kentucky and New Jersey, the Cervus Americanus and fossil elk; in Virginia, the walrus; and in Massachusetts and Louisiana, the fossil horse. The most perfect specimen of mastodon ever found was disinterred in a marl pit near Newburg, N. Y., and is now in Boston, the property of Professor J. C. Warren. From the Big Bone Lick, in Kentucky, it is estimated that one hundred skeletons of mastodon, twenty of the elephant, one of the megalonyx, three of the ox, and two of the elk, have been carried away.

ROCKS AND MINERALS OF ECONOMICAL VALUE IN NORTH AMERICA.

Among the crystalline rocks, such as occupy the greater part of New England, granite is perhaps the most useful, especially if we include gneiss under that term, as is done in popular language. The coast of Massachusetts and Maine furnishes a supply of this rock absolutely inexhaustible; and from thence it is transported to every part of the country accessible by vessels. In Massachusetts large deposits of porphyry occur, the most indestructible and beautiful of all rocks; but in this country it has yet scarcely been used at all. In New England, along its western border, and extending in fact to Alabama, immense deposits of white crystalline marble are found and extensively used; and in other parts of the country marbles of almost every other color are largely wrought, or converted into quicklime. Immense quantities of serpentine occur, also, in connection with the crystalline rocks, though but little used as yet. In the same connection are found large quantities of steatite, one of the most useful of rocks. Of the flagging-stones, perhaps the most elegant is that from Bolton, in Connecticut; and it may be seen in the cities along the whole Atlantic coast. So, also, may the still more enduring rocks from the Hamilton group of the silurian strata, on the banks of the Hudson. The new red sandstone furnishes fine building-stone in Connecticut, in New Jersey and near the city of Washington. That most valuable rock, gypsum, is found in Nova Scotia, in New York, and, in fine, in almost every part of the United States where silurian rocks prevail ; for in this country it is more usually associated with that group, and not with new red sandstone, as in other parts of the world.

The recent explorations of Captain Marcy have made us acquainted with a deposit of gypsum extending from the Canadian river at the ninety-ninth degree of west longitude, south-westerly, at least three hundred and fifty miles, nearly to the Rio Grande, varying in width from fifty to one hundred miles. It is in layers, interstratified with red and yellow clays, sandstone and limestone, and is undoubtedly a tertiary deposit. So I have represented it upon the map, as far as I can do from Captain Marcy's statements. I know of no gypseous deposits to be compared with it, unless it be the Andes, where it is highly metamorphic. It must be of great importance to the south-western portion of our country.

The common salt used in this country is mainly obtained by the evaporation of the waters of the ocean, or, in the interior, from salt springs. In Virginia and in Oregon, however, beds of solid rock-salt have been discovered; and it doubtless exists at the source from whence the salt springs obtain their saline ingredients in other parts of the country. It is an interesting fact, that there occurs near the sources of St. Peter's river, on this side of the Rocky Mountains, a lake, called the Devil's lake, more than forty miles long, whose waters are as salt as the ocean; and beyond the Rocky Mountains, in Utah, is another lake, two hundred miles long, whose waters are saturated with salt. An inexhaustible supply will, of course, be obtained from these lakes, of this most useful substance, when the country is settled.

The extent of the coal-fields of this country has already been given. From the Pennsylvania mines alone, in 1847, not less than two million nine hundred and seventy thousand three hundred and seventy tons were brought to market, at the value of twelve million dollars. In 1851 the amount had increased to four million three hundred and eighty-three thousand seven hundred and thirty tons. In 1845 the amount in all the country was four million four hundred thousand tons, and 1852 it must have been not far from six millions.

Iron is the metal most abundant and most widely diffused in this country, so that no large district exists without rich deposits of it. In Nova Scotia are valuable mines of iron. In the azoic regions of New England, New York, New Jersey and Pennsylvania, the magnetic oxide abounds, especially in the northern part of New York, where the quantity is immense and the quality excellent. Along the western slope of the Green, Hoosac, and some part of the Appalachian Mountains, rich beds of hematite iron abound. In the coal regions of Pennsylvania,

Maryland, and, indeed, almost everywhere, the brown oxide is abundant. But the specular iron ore of Missouri exceeds all other deposits in our country, with perhaps one exception. It is there connected with porphyry, and near to granite and trap. Pilot Knob, five hundred feet high, is partly, and Iron Mountain, three hundred feet high and two miles in circumference, is entirely composed of this ore. This ore has scarcely yet been worked ; and the same may be said of a multitude of other rich deposits of iron in the country, especially that one near Lake Superior, which perhaps exceeds all others in the country in amount, though but little explored.

In the State of Wisconsin, embracing also eight townships in Iowa and ten in Illinois, lies one of the most remarkable lead deposits in the world. Its length is eighty-seven miles and its width fifty-four, covering eighty townships, or two thousand eight hundred and eighty square miles. The rock containing it is the cliff limestone, as described on a previous page. Thus far the lead has been chiefly found in the drift, or in caverns. In 1839 thirty million pounds of lead were smelted from this deposit, and during the eight subsequent years it varied from thirty-two millions to fifty-four millions. It is thought that it will readily yield one hundred and fifty millions annually. Lead ore exists in quantity in other parts of the country, and would be wrought were it not for the remarkable facility of obtaining it in Wisconsin.

Associated with this lead is an abundance of copper ore, not yet so extensively wrought, for want of a market. At Mineral Point, however, more than one million five hundred thousand pounds of copper have been wrought. Zinc also abounds, with the galena, as it does all over the country.

But a still more remarkable region of copper exists in the western part of Michigan, on the south shore of Lake Superior; and it is said also to occur on the northern shore. Native copper has been found in the drift south of this region over an area of several thousand square miles. One mass, on the Ontonagon river, recently transported to Washington, weighs about four tons, and other masses have been dug from the mines weighing fifty tons. Within a few years many companies have been formed, mostly in the eastern cities, for working the copper of Michigan. Several veins have been wrought to a considerable depth, and some of them have been found to be very productive. Native silver, as well as ores of silver, occur in considerable quantity, the first often soldered to the native copper. As the value and productiveness of such deposits are very apt at first to be much overrated, and even geologists cannot form an accurate judgment till extensive explorations have been made, it is difficult as yet to predict how many of the adventurers in these explorations will be repaid, and how many will lose their labor. In Messrs. Foster's and Whitney's "Report on the Copper Lands of Lake Superior," made in

1850, we find nineteen mines described which have been more or less worked. They estimate the amount of copper extracted in 1849 to be twelve hundred tons, and in 1850 they suppose it would rise to two thousand tons.

Ores of copper, as well as native copper, are found in other parts of the United States, associated in the Connecticut valley and in New Jersey, as in Michigan, with red sandstone and trap, exactly alike in their lithological characters, but not probably of the same age. The copper mine at Bristol, in Connecticut, produces very fine ore and in abundance, and is vigorously explored.

Mexico has been celebrated, from its first settlement, as rich in the precious metals. Gold, however, has not been found in great abundance, only about four thousand three hundred and fifteen pounds Troy being annually obtained. But it yields two-thirds as much silver as is found on the whole globe, and ten times as much as the whole of Europe, namely, one million five hundred and forty-one thousand and fifteen pounds Troy. Not less than three thousand mines have been opened. Iron, lead, copper, mercury and tin, are also wrought. Of the latter, Mexico is the only part of our continent where it has been found in quantity large enough to be worked, although discovered in small quantity in New England.

An extensive deposit of gold occurs in the hypozoic or metamorphic strata of the United States, extending, so far as is at present known, from the river Coosa, in Alabama, to the Rappahannock, in Virginia. It is found in veins of porous quartz, traversing gneiss, mica slate, and especially talcose slate. It is usually obtained, however, from the loose soil by washing, and rarely in any country is gold blasted from the rocks, the works having been already performed by atmospheric and aqueous agencies. The same rocks stretch north-easterly through the United States, and occasionally small quantities of gold have been obtained; as at the iron mine in Somerset, Vermont, where I described it many years ago; and recently it has been found in Bridgewater, in the same state, a few miles north of Somerset. The value of the gold sent to the mint of the United States, from our gold deposits, from 1823 to 1836, amounted to four million three hundred and seventy-seven thousand five hundred dollars, about half, it is supposed, of the quantity actually obtained.

The year 1847 was signalized by the discovery of one of the most remarkable and prolific deposits of gold in the world, in Northern California. The region already known to be a gold deposit, — that is, in sand and gravel,— is from four hundred to five hundred miles long, and from forty to fifty miles wide, lying on the western slope of the Sierra Nevada Mountains. The gold is laid open chiefly along the tributaries, from ten to twenty miles apart, which descend from those mountains, and empty into the

Sacramento and San Joaquin rivers; which run, the first south and the other north, and empty into the Bay of San Francisco. These tributary streams are, for the most part dry, during the summer, and in their beds chiefly the gold is washed out. The "dry diggings" are spots where the quartz containing the gold is disintegrated by atmospheric and aqueous agencies.

The prevailing rock in this gold region, as in most others, is talcose slate, abounding in quartz, which in many places would seem to be the principal rock. It is in the quartz that the gold occurs. There are also deposits of miocene sandstone above the quartz and the slate; but these have no connection, probably, with the gold. Extinct volcanoes exist, also, in the Sacramento valley; as the *Sacramento Bute*, one thousand eighteen hundred feet high, composed of porphyry and trachytic porphyry. Igneous rocks also occur in the Sierra Nevada Mountains. In short, this gold region corresponds almost exactly with that in the Ural Mountains, and in the Altai chain, and many other parts of eastern and northern Asia; and the inquiry can hardly fail to force itself upon one's attention, whether the Ural Mountains may not be the western, and the Sierra Nevada, or perhaps the Rocky Mountains, the eastern margin of a vast gold deposit, which has already attracted the attention of the whole world.

Hon. T. Butler King, in his report to the United States government respecting California, in March,

1850, estimates the amount of gold dug there in 1848 and 1849 at forty millions of dollars. And there is every reason to suppose that the quantity obtained will be increasing for a long time. If this estimate is not much beyond the truth, such an influx of this precious metal, taken in connection with the large and increasing quantity from the Russian mines, must materially affect the value of the circulating medium in all countries, and render it possible to employ gold more extensively in the arts; and, since the discovery of the electrotype process, we may hope that a large part of metallic articles may ere long be coated with gold, and thus secured in a great measure from oxidation.

Nearly all the gold of commerce has for a long time been obtained from Asiatic Russia, Brazil, Transylvania, Africa, the East Indian Islands, and the Southern States of this republic. The annual supply from these sources has been about eighty thousand pounds, in value not far from twenty million dollars, or a little more than half the amount dug in California in 1849. A writer in the *London Athenæum* estimates the supply for 1851 at one hundred million dollars, of which he thinks seventy-five million dollars will come from California. Probably this did not exceed the amount, and we may safely set down the future supply at not less than one hundred million dollars. Platinum, too, occurs with the gold there, as well as extensive mines of quicksilver, one of which, in 1851, yielded about one million two hundred thousand pounds.

The gems or precious stones of this country are so seldom cut into jewelry, that it is difficult to say how valuable our localities may be. Most of the gems, however, occur on this continent. The diamond has been found in North Carolina, as well as in California. The blue sapphire occurs in New York, New Jersey, Connecticut and North Carolina, especially in large crystalline masses; the chrysoberyl, at Haddam, in Connecticut, and in Saratoga County, New York; the spinel, in New York and Massachusetts; the topaz, at Monroe, in Connecticut, in large quantity, but usually massive; the beryl, of great beauty, at Royalston, Massachusetts, and at Haddam, in Connecticut, as well as at many other places in New England (*e. g.*, Ackworth, N. H., Barre and Goshen, Massachusetts); the zircon, or hyacinth, in New York, Canada and North Carolina; the garnet, almost everywhere in the crystalline rocks, and the pyrope, especially, of great beauty, in Sturbridge, Massachusetts; the cinnamon stone, at Carlisle, Massachusetts; the green and red tourmaline, of great beauty, at Paris, Maine, and Chesterfield, Massachusetts; limpid quartz, at Little Falls and Fairfield, New York; the smoky quartz, in various places; the amethyst, in Nova Scotia, Bristol, Rhode Island, &c.; jasper, hornstone and chalcedony, in the various trap ranges, especially along the south shore of Lake Superior, forming agates; chrysoprase, at New Fane, Vermont ; iolite, at Haddam, Connecticut, and Brimfield, Massachusetts; adularia, of various colors, in Brimfield, &c.; labordorite and hypersthene, in Essex County, New York ; cyanite, at Chesterfield, Massachusetts, &c.; fluor spar, in the northern part of New York ; satin spar, at various places ; precious serpentine, at Newbury, Massachusetts ; red oxide of titanium, at Middletown, Connecticut, and Windsor and Barre, Massachusetts ; and manganese spar, at Cummington, Massachusetts.

CONCLUSION.

From the brief summary that has now been given of the geology of the globe, some general inferences may be derived, such as the following :

I. We find that in nearly all the great mountain chains on the globe the same leading phenomena exist ; for example,

1. The central axis of the chain is composed of crystalline rocks, stratified and unstratified.

2. The flanks of the mountains are usually covered by stratified deposits, showing a mechanical origin and structure, often traversed by dikes of unstratified rocks, nearly compact or vesicular ; *e.g.*, greenstone, basalt, &c.

3. The plains are generally composed of rocks derived from the erosion of older rocks, and only imperfectly consolidated, the strata being more or less horizontal.

4. Upon and among the newest stratified deposits we find the most recent unstratified rocks; *e. g.*, trachyte and lava; wherever, indeed, volcanic action has operated during the tertiary or post-pliocene period.

5. The crystalline schists, as well as the unstratified rocks, are destitute of organic remains; but they are found in all other rocks.

6. The surface has been subject to powerful erosive agencies, atmospheric and aqueous, as is proved by the fact that the newer mechanical rocks are composed of the ruins of the older.

II. The ocean has stood above our present continents, and for a long period, as is proved by the fact that the organic remains of the fossiliferous rocks are mostly marine.

III. The disturbed and tilted condition of the older rocks, forming the axis of mountains, shows that they have been lifted up by lateral pressure or internal force.

IV. The phenomena of terraces and ridges of sand and gravel render it probable that the desiccation of our continents has been partly the result of a subsidence of the waters of the oceans.

V. It is probable that continents once above the waters may now occupy the beds of our present oceans.

VI. The geological structure of one continent, or of one grand mountain chain, is a type of the geology of the globe.

VII. Hence the leading principles of geology may be considered as established ; and, though explorations in regions but little known may bring to light new and interesting facts, we have no reason to suppose they will ever essentially alter any of the leading principles of the science. Every continent has now been sufficiently explored to make it sure that the same great laws were concerned in the geological structure of them all.

INDEX.

Telerpeton Elginense. *From the Old Red Sandstone.* B.

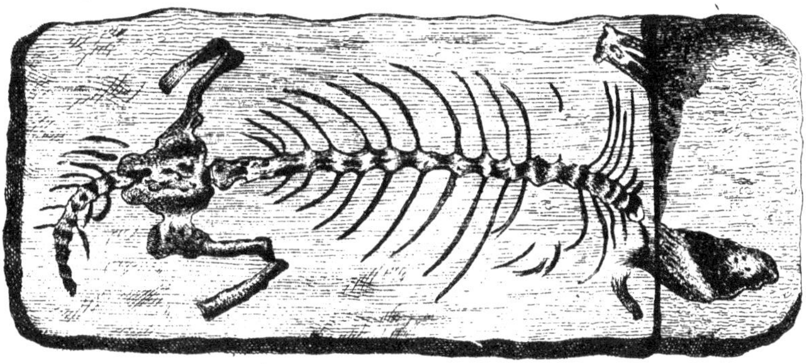

Chelonian Footmarks. *From Old Red Sandstone.* A.

3.

2.

1.

7.

6.

4.

8.

5.

W.C. SHARP'S LITH. BOSTON.

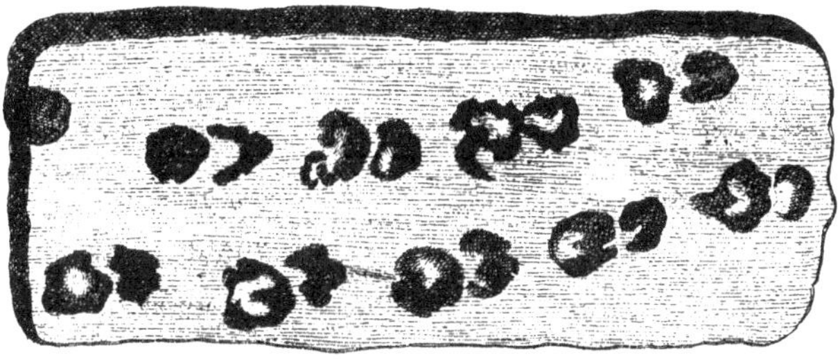

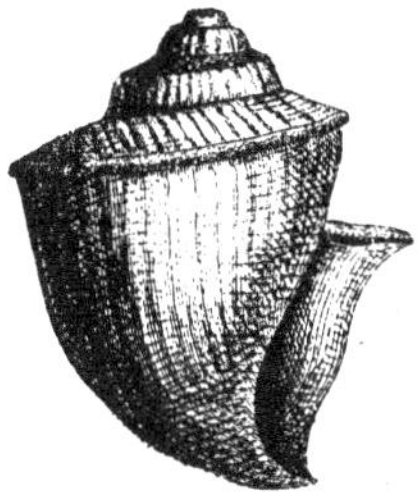

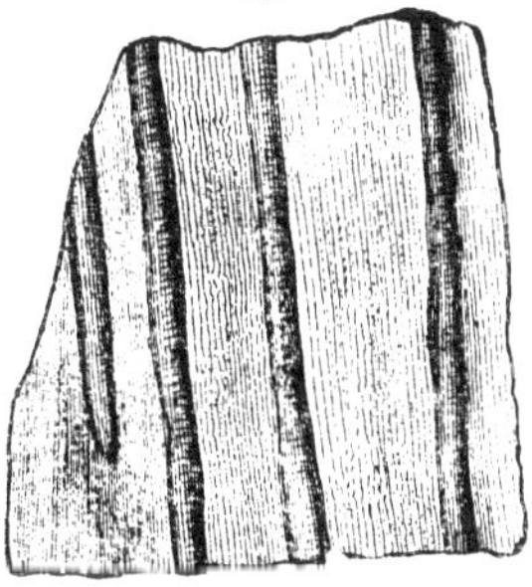

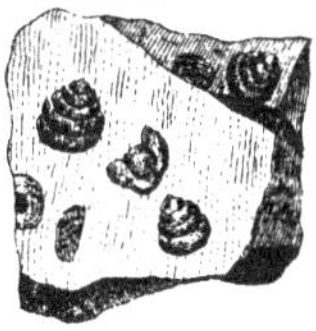

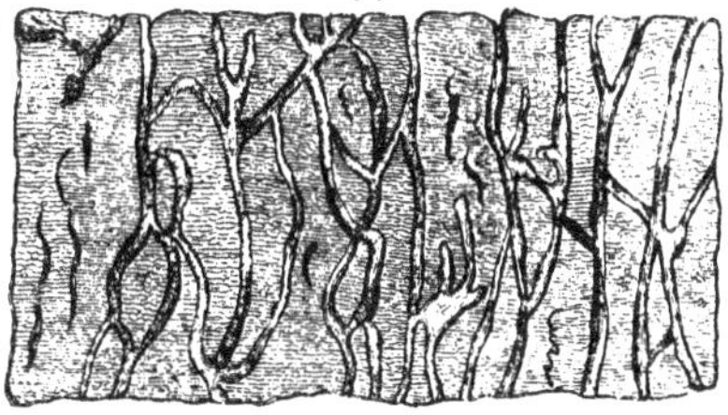

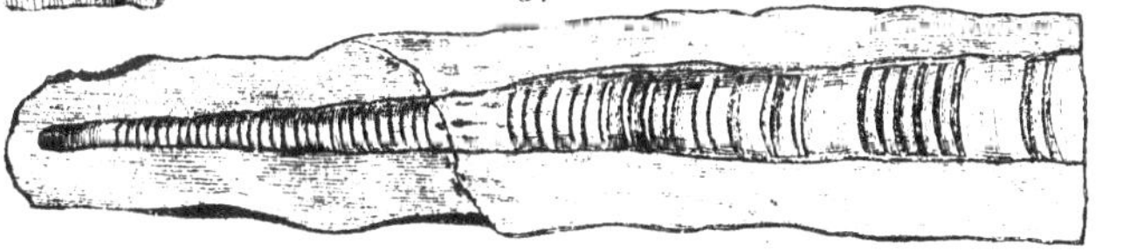

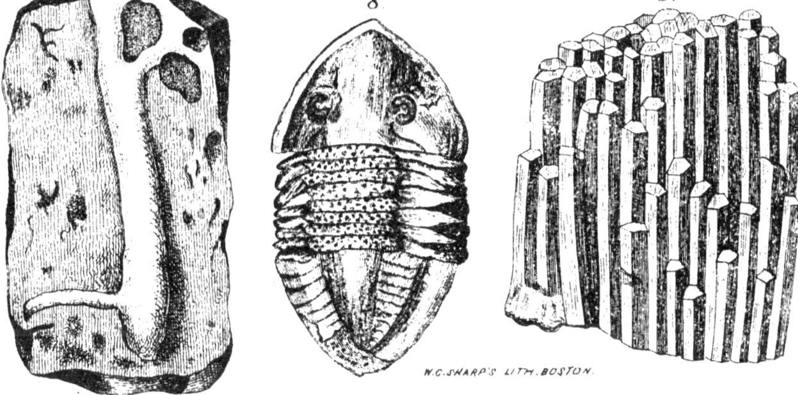

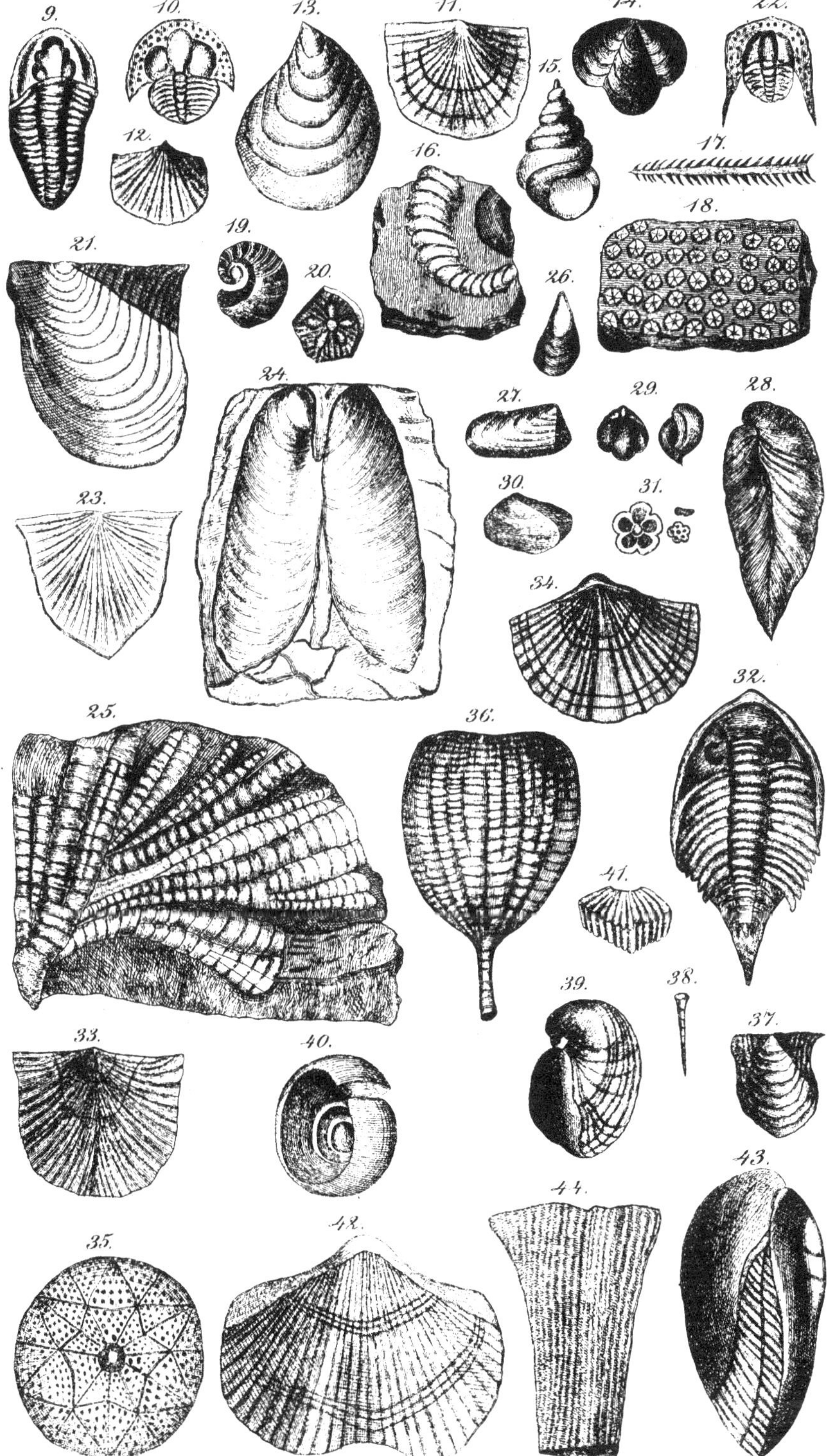
9.
10.
13.
11.
14.
22.
15.
12.
16.
17.
18.
19.
21.
20.
26.
24.
27.
29.
28.
30.
31.
23.
34.
32.
25.
36.
41.
39.
38.
37.
33.
40.
43.
44.
35.
42.

45.
46.
47.
49.
48.
50.
51.
53.
52.
55.
54.
60.
57.
58.
66.
61.
59.
63.
62.
64.
65.
56.

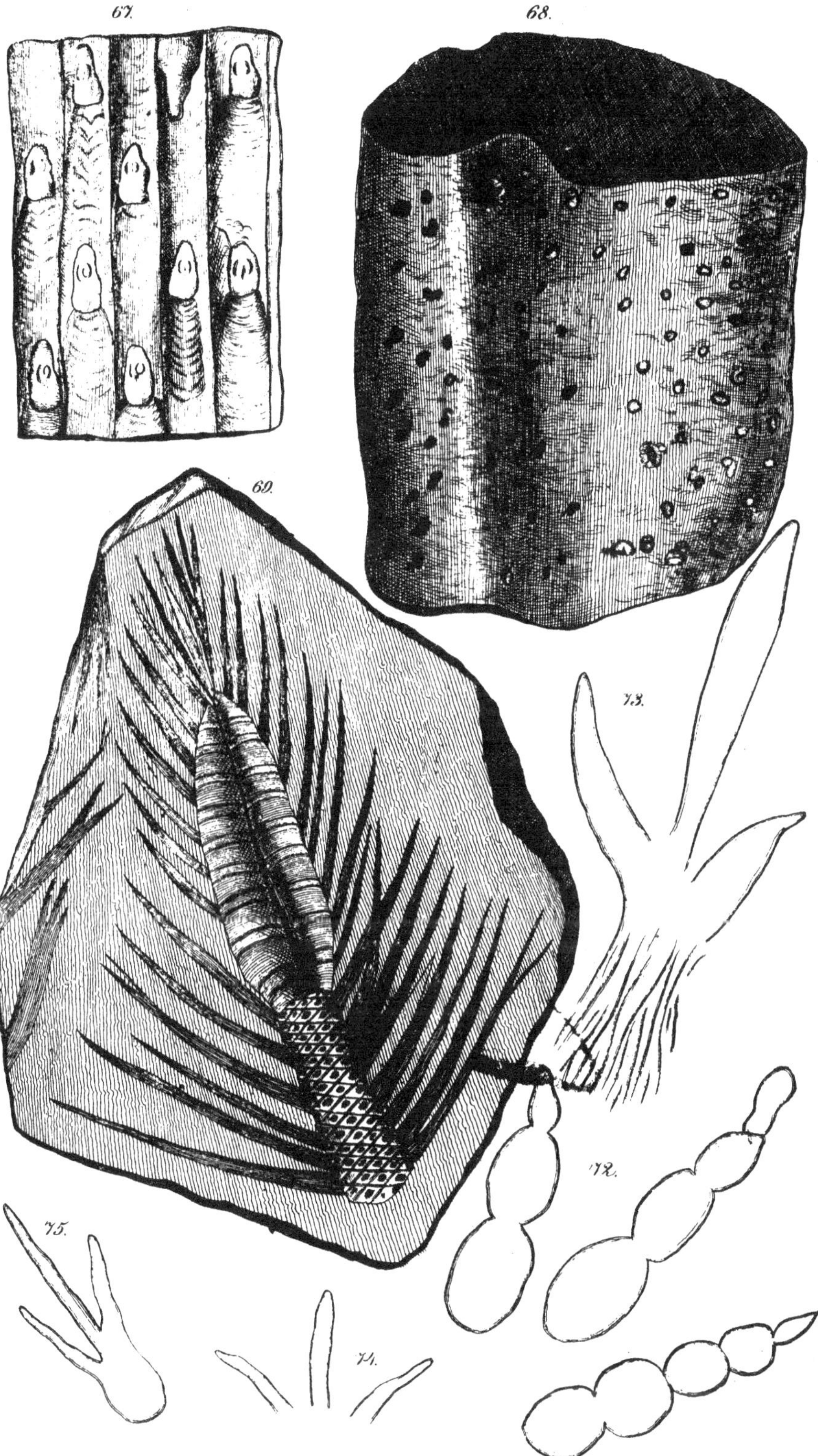

For Fig. 70 & 71 see Plate V.

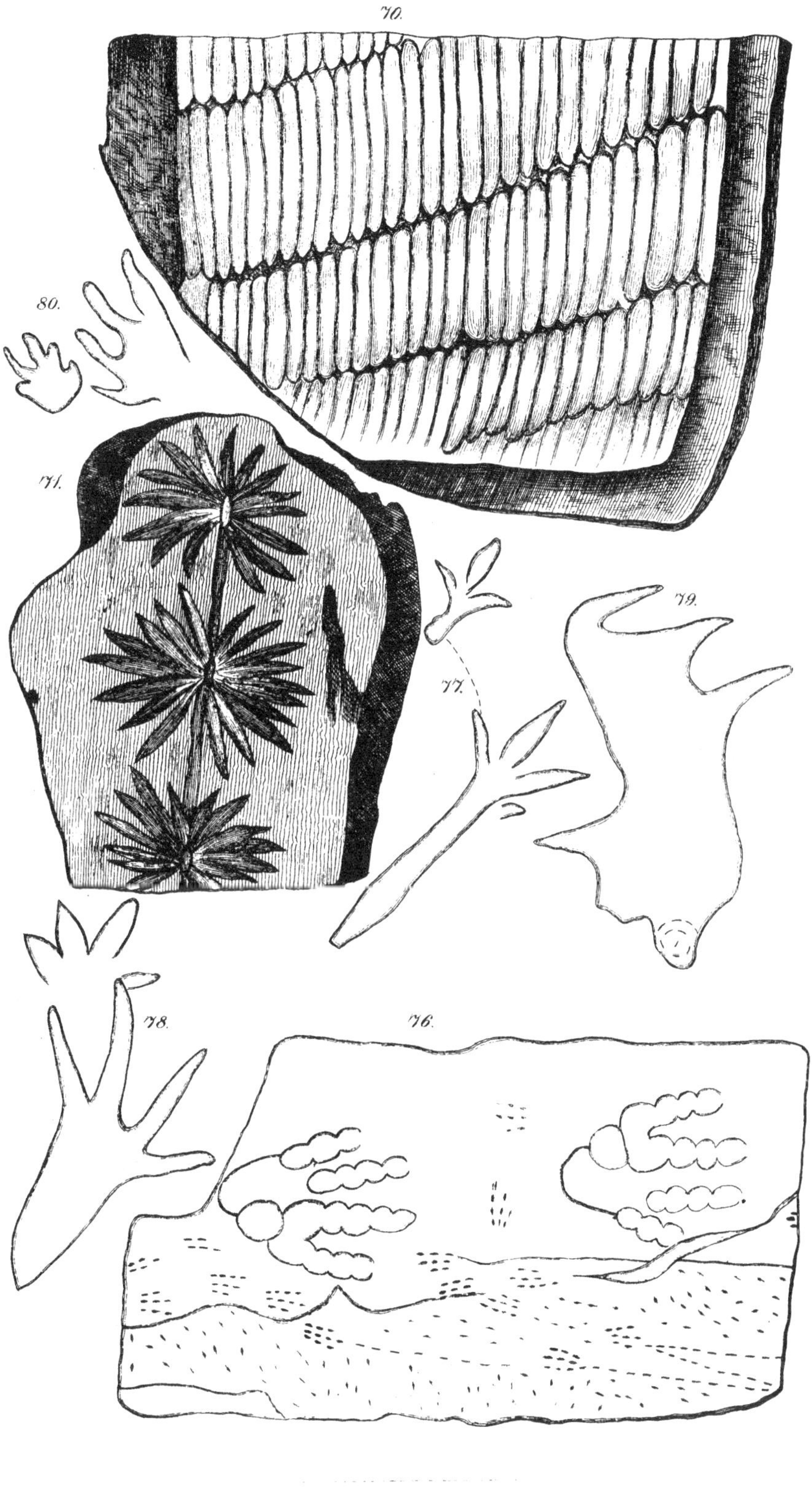

70.
80.
71.
79.
77.
78.
76.

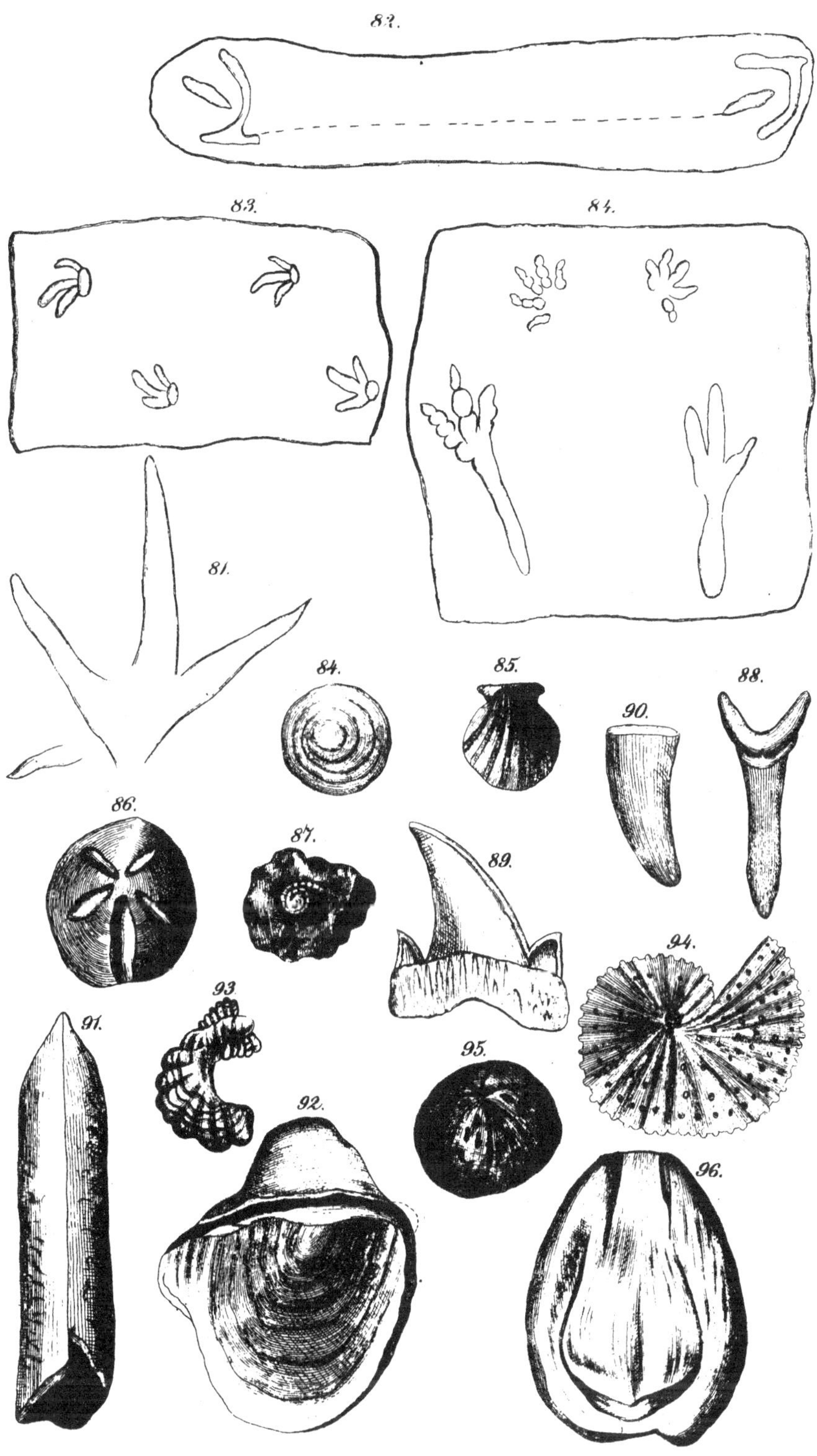
82.
83.
84.
81.
84.
85.
90.
88.
86.
87.
89.
94.
93
91.
95.
92.
96.

www.ingramcontent.com/pod-product-compliance
Lightning Source LLC
LaVergne TN
LVHW021412110826
845150LV00007B/1886

* 9 7 8 1 4 2 5 5 1 0 9 5 4 *